DICKINSON AREA PUBLIC LIBRARY
3 3121 00131 3142

D1006234

*Including essential Country Woman recipes

Gwen Petersen

Voyageur Press

First published in 2007 by Voyageur Press, an imprint of MBI Publishing Company LLC, Galtier Plaza, Suite 200, 380 Jackson Street, St. Paul, MN 55101 USA

MBI Publishing Company titles are also available at discounts in bulk quantity for industrial or sales-promotional use. For details write to Special Sales Manager at MBI Publishing Company, Galtier Plaza, Suite 200, 380 Jackson Street, St. Paul, MN 55101 USA

Editor: Margret Aldrich

Designer: Sara Holle

Printed in United States of America

Library of Congress Cataloging-in-Publication Data

Petersen, Gwen.

How to shovel manure and other life lessons for the country woman / Gwen Petersen.

p. cm.

ISBN-13: 978-0-7603-2862-0 (hardbound w/ jacket)

ISBN-10: 0-7603-2862-5 (hardbound w/ jacket)

1. Farmers' spouses–Handbooks, manuals, etc. 2. Farm life–Handbooks, manuals, etc. I. Title.

S518.P467 2007

630.92–dc22

[B]

2006100836

DEDICATION

To all ranch and farm women and men who work the soil to feed and clothe our nation. To all my wild woman and cowgal pals . . . you know who you are. And to my beloved horses.

PARTS AND INNARDS

PART THREE

ESSENTIAL COUNTRY WOMAN RECIPES

PREFACE

Over the decades of rural living, Country Women have been helpmeets to their men. With remarkable courage, incredible stamina, and splendid insouciance to the rawer facts of nature, the Country Woman has shaped the nest surrounding that hardworkin', cow-wrasslin', bull-throwin' son-of-a-farmer. Take away the Country Woman and it's like throwing away the keys to the pickup: Nothing gets going.

This book is a handy guide and aid to those choosing to spend their lives as Country Women. It contains really helpful, practical, useful tips and hints to keep the Country Woman from going berserk. (Reserve going berserk for special occasions.) Because farm and ranch life is necessarily divided and marked by the seasons of the year, the manual will attempt to follow a similar pattern. Our hope is to provide an understanding reference volume for the dauntless Country Woman, couched in language developed in secret women's conclaves* (see footnote), not necessarily comprehensible to men.

Any Country Woman could write a book about country life. It has all the elements of good fiction: excitement (you hope that young colt won't dump you), tension (if you do get dumped, you hope nothing on your personal self breaks), danger (is that a rattlesnake curled under the tractor?), heroes and heroines (that would be your Dearly Beloved, yourself, your family, and your favorite sheepdog), and villains (that would be government regulations, mean roosters, wolves, coyotes, other uninvited varmints, and sometimes the plumbing).

But country life isn't fiction. It's as real as the sun rising each morning, as genuine as snow on mountaintops, as basic as breathing. Country life is the story of land.

*Conclaves: Referred to by Country Men as "gab-sessions."

Nurturing the earth is the heart of country life. In a rural culture, ranchers and farmers are the first stewards of land that produces the food and fiber that feed and clothe the nation. As you watch and work with the magic of changing seasons, your creed as a Country Woman is simple and true: *Take care of the land and the land will take care of you.*

COUNTRY LIFE: A TO Z

Agriculture's country rhythms city folk don't know
Blatting lambs and bawling calves and colts that whicker low
Counting critters, checking heifers, calving out the cows
Dally ropes and playful doggies, time to farrow sows
Ear-ing down a snorty bronc and getting set to brand
Finding strays and feeding bums and fencing 'cross the land
Grasses, grains, and gardens flourish—rich with food to eat
Hay and horses, hens and eggs, and bees make honey sweet
Irrigating, shoveling mud, and hauling heavy dams
Jug the ewes in early spring, assist the birth of lambs
Kittens mew in plaintive voices in the milking barn
Lambs grow fat and ewes grow wool to make the fine-spun yarn
Mow the hay and harvest grain in summer until fall
Nestled calf behind a bush awaits his mamma's call
Oysters cut from scared bull calves sure make a gourmet treat
Powder River, let 'er buck! The cowboy keeps his seat?
Quarter horses, Appaloose, Mustangs in the West
Riding ranges, rural fun, the country is the best
Sheep are trailed to mountain grass, coyotes stalk a lunch
Trail the woolies home in fall, the bucks now join the bunch
Underwear beneath his Wranglers warm a puncher's hide
Vegetation chokes the coulees where the cowboys ride
Wild and woolly, full o' fleas and when the work's all done
X the dates till Xmas comes, the days are short of sun
Yoke the teams when snow is deep, the livestock must be fed
Zany fun at New Year's dance, oh how this year has sped!

PART ONE

Springtime on the Ranch and Farm

Spring: That time of year when the earth thaws and releases the frozen moisture, creating a thick, dark gumbo usually tracked across the kitchen floor in artful patterns. The last of the hay for feeding the cows is running low and your man is worried for fear he'll be forced to buy more. He worries EVERY year.

SECTION ONE

Calving the Heifers

Wherein one learns about the midnight duties required of a good Country Woman, and how to serve as midwife at the birth of a calf.

When the two-year-old heifers are heavy with calf, your man makes many trips to the north pasture to check on possible new arrivals. Checking the heifers is an around-the-clock job in which you get to share. Your turn is the midnight and four a.m. tour. Bundle up to the eyebrows and arrange earmuffs* (see footnote). Stuff your feet into your husband's old overshoes (for some reason one of your own boots has temporarily disappeared). Pick up the large flashlight with one mittened hand. Huddle deep into your muffler and track through the snow and the dark to the calving shed. Carefully check each and every heifer confined there. To check a heifer means shining your light on the delivery area for signs of an imminent grand opening. Usually the critters turn to stare at you staring at them. Then check each heifer just once more because you know you'll be chewed out if you miss one and she calves behind your back! Thankfully none of the girls seem ready to start yet. You pray they'll all wait for your husband.

*The Country Man NEVER wears earmuffs. In fact, it has to be twenty below before he'll lower the earflaps on his Scotch cap.

Stagger back to the house and climb once more out of all those clothes. As you fall gratefully into bed, your man will cease snoring, open one eye and ask, "Everything all right?" Receiving an affirmative, he's instantly unconscious again.

However, not all your nighttime trips to the shed are so uneventful. More likely one of the darlings has begun the business of giving birth. Draw a deep breath and wonder what was so bad about secretarial work. Then inspect the situation. If the heifer is in labor, but nothing is poking out the back end, you're in luck. Tell the heifer to cross her legs, then as rapidly as mud or snow permits, stumble to the house where you awaken the cowman. What a relief to turn the problem over to him! Sometimes though, you'll discover a heifer down, moaning, and with two little feet protruding from her rear. She's in trouble. Resist the urge to panic and run for house and husband. Time is of the essence, he always more or less says. If the cow seems disinclined to get up, chances are she'll stay there while you work, you hope. Bravely shed your mittens, plunge your hands into the bucket of soapy water and disinfectant (you've kept this pushed down to the lid in the oat bin), and approach the young mother. Timing your efforts with the contractions, begin a steady pull on those two protruding legs. With any kind of luck you'll be strong enough to pop the little guy out of there. Recommended pull position is to sit spraddle-legged on the floor, brace a foot on either haunch of mamma, and pull like mad whenever she contracts. But sometimes nothing happens. Mamma is trying, you're trying, but the calf remains stuck. By now you're scared it'll suffocate. Should you race for the house and husband? Should you get the calf pullers? (A loop affair of chain on the end of a stick with a sort of ratchet thingy. You can slip the loop around the calf's legs up inside the mother and thereby achieve a much stouter grip and stronger pull.)

One thing you don't notice anymore is the cold. Your eyes are a little blurred though, from the tears of desperation. Rinsing your hands once again in the disinfectant, ease a hand into mamma. To your horror you discover it's a BREECH birth! Dear Lord, now you're sure the calf'll be dead when it finally arrives. Grab the calf pullers, slosh 'em in the disinfectant, and ease the loop over those legs. Assume your spraddle-footed position. The heifer contracts mightily, you pull fiercely, teeth gritted and eyes squinted. You've just pleaded once again with the Almighty, when with a sudden slurpy release, the calf squishes forth. Whereupon you roll backwards into the fresh cowpie behind you. But clamber to your feet instantly and hover over that calf. He's alive! Work his little legs back and forth and massage his sides. Finally, he lifts his head and blinks. By this time you're crooning a loving refrain to him as you hoist him up so you can iodine his navel.

Before you leave, make sure mamma is on her feet. Push the baby right under her nose so she'll get on with the job of licking him off* (see footnote). Gather your tools, put away the bucket of disinfectant, find your mittens, and stumble—exhausted, but proud—for the house.

"Everything all right?" asks your loving man as you wearily shed your smelly clothes.

"Just fine," you say. "That number 35 heifer had a nice big calf."

"Oh? Any trouble?"

"Well, a little," you say. "Mother and child doing well, but I DID have to. . . ."

"Heifer or bull calf?" your beloved mumbles.

"A nice big bull calf," you respond warmly. "But he was backwards and I had to use the calf pullers and . . ." a wheezing snore interrupts your proud recital. Philosophically, you relax contentedly alongside the snorer, murmuring, "No—no trouble, dear, no trouble at all."

The brockle-faced cow had a fit
When her calf caught hold of a tit.
"Oh, surely this tot'll
Survive on the bottle,"
Said she, "Not on me, cause I quit!"

*Sometimes mamma is already on her feet and making menacing movements towards you. Do not stand there arguing about child-care methods. Exit quickly.

SECTION TWO
Chickens

Wherein one learns about chicken manure, chicken-house cleaning, egg gathering, setting hens, butchering roosters, chicken plucking, and chick hatching.

Chickens are those remarkable feathered creatures that come in a variety of shapes, colors, and sizes. But no matter how varied, once a chicken—always a chicken.

Generally, the man of the ranch or farm hates chickens. Probably due to chicken hysterics. A chicken is an inconsistent creature that squawks, flies up onto the shop roof, and nests in King-the-Collie's doghouse. At egg-gathering time, you must remember to check the doghouse before King carries off the egg. (King hates those eggs making lumps in his straw.) Egg gathering is not exactly an art. Anyone can do it. Even Number Two son, age four and a half, can do it. However, gloves are recommended if you're chicken about sharp beaks. Occasionally you will meet a truly mean chicken—usually a rooster. The treatment of choice is to plan stewed rooster for Sunday dinner. Especially if it's a large, evil bird inclined to attack defenseless small children, cats, and dogs. In case of chicken attack when you are carrying eggs, hold the precious cargo aloft, leap high, and karate-kick your foot at the fowl.

Number Two son learns to develop a sharp eye and instant reflexes when dealing with the meanest rooster. Try not to scream or panic when you glance out the kitchen window to observe Roger the Rooster sailing through the air, talons extended, to spike poor Second Son on his small behind, knocking him flat. With an ear-splitting scream, small Son quickly hunches to his knees, scrambles to his feet, and beats it for the house. All this he does with both fists doubled and held straight up over his head. The reason for this peculiar posture becomes apparent as he stalks into the house, hands still poked up over his head. He marches to egg basket, carefully lowers his arms, and deposits two eggs therein.

Frowning horribly, he about-faces, stomps back out the door, pausing only to grab the broom. Roger the Rooster dodges for cover as a vengeful small boy tries to beat the fowl's brains out.

On the day you choose to butcher chickens, secure a short-handled axe, a sharp machete, or a cleaver* (see footnote). Keep the chicken-house

*If you can possibly secure your husband's services to wield the implement, do so. Or corral the neighbor, or anyone else available.

door closed (with chickens inside, of course). Slowly and quietly sneak in, closing the door behind you. Then crouch and cover your head as thirty-seven hysterical chickens fly past and over. Grab one by the legs, ascertain that you wish it to be sacrificed, back out the door, and close it again. Fling the chicken on the block (hang onto the feet) and grasp your decapitating tool in your other hand. A flung chicken automatically will extend its neck. Whereupon you guillotine it. You have to be quick and have a sure aim. For those of us not blessed with surety and speed, a more efficient and certain method is to acquire a stump and pound in two big-headed nails side by side, leaving them protruding about two inches—just wide enough apart for a chicken neck. Pulling on the chicken feet causes the bird to lengthen its neck, thus making it making it easier to whack the correct spot. Try not to think or imagine too much as the body runs and flops across the ground while the head lies there gasping and blinking. Simply go after another chicken.

When you've beheaded about six—quit. That's going to be enough plucking for anybody's day. Gather up the feathered bodies by the feet—three in each hand—and hold them out away from you so the blood doesn't spatter on your legs. Then run like mad for the plucking spot.

On the kitchen stove you've got a huge pot of water boiling. Drop chickens on ground. Fetch a pail of the steaming water. Dunk one chicken at a time into water. (Adding detergent—but not too much—to the water seems to make for easier plucking.) Slosh chicken about for a few seconds. Pull out and start plucking. Pull feathers as fast as you can* (see footnote).

Although you can pull the big feathers from several fowl in just no time, it's those dinky pin feathers that ruin your record. Pin feathers are an important subject when discussing chicken plucking. It takes forever to get them all out. When you've achieved a pin-feather-free fowl, it's time to singe your bird.

Chicken singeing is a must. The flame sears away chicken hairs left after plucking. Wad and twist newspaper, or squash and twist a paper sack. Strike a match to the end of your makeshift torch. Hold chicken and quickly pass the burning paper torch all over the bird. This method scorches off chicken hairs but is apt to scorch off your skin as well.

Alternate method: A small flat tuna can half filled with alcohol, lacquer thinner, charcoal lighter fluid, or—if desperate—gin, bourbon, or brandy will

*Speedy chicken pluckers rate almost as high as home-baked bread artists.

flame nicely when touched with a match. This allows you to hold the chicken with both hands, thus insuring uniform singeing. CAUTION: If you use gin, bourbon, or brandy, NEVER disclose this fact to your husband.

After singeing, you're ready to gut the bird. Take up your nice, freshly sharpened (by your dear husband) butcher knife and go at it your favorite way. (Actual gloppy details are just plain yucky. If you don't know how to do it, ask your neighbor. Better yet, prevail upon your bridegroom to do the deed.) After the birds are gutted and cleaned, cut them up into the usual pieces (unless you want to leave them whole for stewing or baking) and drop into a bucket of cold water to rinse. Remove, drain, and wrap 'em in butcher paper or drop in a plastic bag. Stick them all in that wonderful invention, the freezer.

If you feel there ought to be an easier way to reduce a chicken to the eating stage, there is. One way is to get someone else to do the butchering. Another is to skin the chicken rather than pluck the blamed thing.

However, your man will insist that the skin is the best part to eat! Listen to this plea for one or two birds. Skin the rest.

Chicken-skinning how-to: Pound two nails side by side on the outside of the woodshed about eye level. Tie the chicken's legs, one to each nail with a piece of baling string. Start the skinning by slitting down each leg. If you get really competent at this activity, you can peel that whole skin off in no time. Remember, if you skin, you don't have to singe.

At times it becomes necessary to clean the chicken house of all that accumulated chicken guano* (see footnote). A foul chore, but necessary only a few times a year. Pay no attention to those who do it more often. The tools of choice are the pitchfork, the grain shovel, and a spade. What you do with these implements should be obvious. A wheelbarrow may be needed to transport the stuff. That treasure the chickens have been depositing for months must now be gathered up and redeposited on your garden. Eventually you'll be through toting and carrying. Then creosote the interior of the chicken house till you gasp for air. Put fresh straw in the laying boxes and on the floor. You are finished. You now resolve to allocate this chore to your oldest son since it is well known that shoveling chicken excrement is character building. Many politicians started out as common chicken shovelers.

Hens are odd creatures. They yearn to become mothers. Anybody's egg will do. You'll discover hens in nooks and crannies all over the farm snuggled

*SMALL chicken houses are recommended. Shoveling chicken manure is easier if it's concentrated in a small area.

down tightly over their eggs, glaring balefully. There's something unnerving about the unblinking round-eyed stare of a setting hen. You can tell she's trying to set when she screeches wildly but won't LEAVE as you reach under for eggs. The peck-holes on the back of your hands are also indications of a needy hen.

Pay no attention to when the hens want to set. Make them wait till spring so the little ones will have the warm summer to grow and prosper. Twenty-one days after you put a batch of fertile eggs under a hen, magic happens. Fuzzy little chirping creatures can be seen peeking from under mamma, who will fight to the death for her chicks. You've conscientiously kept mamma fed and watered in her personal cage, checked the eggs a time or two, suffering peck wounds, but still you can't believe that's all there is to it. You just wish it had been that easy for you when your last child was born.

Keep mamma and chicks separated from the rest of the flock for a few days. When the chicks are strong enough, Mother Hen will parade the group out the door and into the pasture, clucking and scratching. The investigating house cat, sneaking too near, will receive a furious thrust from sharp talons. The smart cat needs only one lesson.

There are no absolutes when dealing with chickens. As an observant Country Wife, you will enjoy chicken antics and always name the roosters after your husband's friends.

"Oh, good," said the dumb little chickens
"We hear we're the very best pickin's."
But when they were chosen,
They ended up frozen,
And later became finger-lickin's.

The people who dearly love chicken soup
Have probably not cleaned a chicken coop.
Do you think they would savor it
If they knew their favorite
Was started among all that chicken poop?

SECTION THREE

Pigs and Pigging

Wherein one learns that pigs require tender loving care and that it is the sympathetic Country Woman who can soothe a troubled sow. Pig care is considered a Womanly Art.

Pigs are those fabled, peculiar-looking creatures credited with a variety of weird habits. Mostly, the fables are exaggerations of the truth by frightened or uninformed persons. As with people, there are chauvinist pigs and there are fine swine.

The lord and master of the family-size ranch or farm usually ignores pigs. He does not form attachments nor become emotionally involved. The Country Woman, on the other hand, can't think of anything cuter than a baby piglet.

Grown-up pigs will root through your fence and eat your garden. They'll root great holes in the irrigated pasture looking for angleworms. A pig will run amuck in the cornfield, chomping ears and tromping stalks. In fact, pigs force their way into or out of anywhere they want. It takes understanding to really love swine.

Pigs actually DO have "little piggy eyes" and "eat like hogs." Their manners are atrocious. The "boss" sow will steal food from the smaller pigs and they in turn chew out younger ones. There's no such thing as concern or fair play in the heart of a sow. Still, she's a friendly creature that loves to have you scratch her back, rub her stomach, or oil her itchy places. She'll come a-running when she hears you or sees you approaching. She'll hang around and talk to you when you are cleaning the sheds. She will also run over the top of you to get at a kernel of grain.

Some people read poetry about the "Children's Hour." You, as a Country Woman, can recite a good deal of blank verse about pigs. Your "swine song," as your husband puts it, includes admiration along with large doses of exasperation. When your favorite sow produces fourteen healthy babies, it's a fine feeling. When she mashes all of her five hundred pounds on your instep and jams her head into the bucket you're carrying, almost separating your arm from the shoulder, a different sort of feeling pertains.

But the Country Woman has the touch with pigs. Women can talk pig. Communication on the deepest level exists between woman and sow. Between man and pig, only suspicion and cold tolerance exist.

At pig-feeding time (7:30 a.m. and 5:00 p.m.), you will cease whatever you're doing and slip into your striped, one-piece, ultra-charming coveralls. First, though, go to the bathroom because a nature call while wearing the things can be awkward. Grab your gloves and don't forget your hat. Recommended hat styles are baseball-cap type or the sailor-hat style. These have enough brim to shade the eyes but not so much you'll keep knocking the blame thing off as you crawl in and out of pig pens.

On a family farm, sow pigs run loose in the south-forty pasture that parallels the river. All swine love to wallow in mud and some enjoy swimming. They can't do the crawl or a backstroke; they merely dog-paddle (pig-paddle?). When farrowing time arrives (three months, three weeks, and three days after their weekend with Beauregard, the Boar) the sows are shut up in their individual apartments in the pigshed.

As you start for the shed, begin calling the pigs. There are several popular styles of pig calling. Choose one or make up your own. Pigs don't care. A favorite of the Country Woman is "Come pig!" announced in a coaxing soprano. The Country Husband generally favors the deep-throated roar of no particular enunciation, but immensely effective. Out of the brush will come galloping pigs, cattle, horses, and all flavors of farm animals. One feels sure elephants would also respond.

You commence to the sheds, pigs, and dogs loping along in a loose caravan behind you. It is best to arm yourself with a stout stick when carrying the grain. The stick is employed as a deterrent to enthusiastic swine. Twelve five-hundred-pound sows jostling you can lead to a loss of dignity. A sow attempting to dart between your legs is bound to be upsetting. Screams and threats on your part make no impression. But a stout clout on the snout with a stick will cause her to back away with an indignant squeal.

As you march rapidly from granary to feed stalls, wave the stick behind you giving the effect of a temporary tail. WARNING: Do not stop your forward progression! To pause is to be inundated by a dozen squealing sows, all of whom will try to dive into your grain bucket.

Each pig will choose her own feed stall* (see footnote), charge in, squeal, rear up on hind legs, and hang over the front edge of the stall, snorting and snuffling and generally raising an awful din. As you go down the row of stalls dropping scoops of grain into troughs, a lovely, peaceful

*The stalls are built in a long row—sort of like starting gates at a race track, but with feed troughs at the head end. A crossbar at the back keeps the pigs locked in. An alleyway in front allows you to dish up the sows' dinners one by one into individual feedboxes.

quiet ensues, except for the unmannerly slurp-slurping. Lock the boss sow into her feed stall. Otherwise she gulps down her grain, scoots out backwards, and tries to ram her way into the neighboring stall already filled with a fine swine. If the occupant doesn't instantly scramble out, the boss sow bites her on the rump, sending the poor thing into screeching hysterics.

As soon as the pigs finish their meal, they will beg and plead noisily for one more nibble of ration. But never listen to a seductive sow. If she gets too fat, she does not do well at farrowing time* (see footnote).

Everything you've heard about "slopping hogs" is true. In a bucket handy to the sink, save all table scraps, peelings, apple cores, onion skins, carrot tops—anything except coffee grounds and peach pits. Once a day, haul this goulash to the pigs. Or dump it into the separated milk your husband puts in the milk can out by the back door and haul both milk and scraps at once. This divine mixture is called "slop" and the pigs adore it.

Farrowing time is an important event on the farm or ranch. In December and in June, the Lone Ranger (the boar) has done his duty by the girls. In spring and in fall, three months, three weeks, and three days after conception, it's time for the ladies' confinements. You've cleaned the pens in the farrowing shed (pig-shed cleaning is dealt with in another section), sprayed and disinfected everything in sight, and strewn fresh oat straw in the farrowing beds and creep feeders** (see footnote). Keep a watchful eye on the pregnant ladies and two or three days before the sow or gilt is due, shut her in her pen. You hope she will decide she likes her pen and wants to have her babies there. Caution: Don't wait too long to shut her up! A sow who decides to nest out in the timber can be highly resistive to alternate suggestions. That's if you can find her at all.

You have already prepared a medicine box containing clean rags, scissors, iodine, alcohol, and a magazine. The reading material is used to while away the time. Be sure it's light reading as concentration is difficult under a single hanging light bulb while sitting beside a farrowing sow. She won't care if you read to her.

It will usually be late when the sow pigs. One a.m. is a favorite hour. However, several hours before she lies down for her grand opening, a sow will "make her nest," which means she chews all the straw into little pieces, eats great chunks of wood from the sides of her pen, and, several times, tries

*Farrowing: When the pregnant sows deliver their multiple blessed events.

**Creep feeder: A narrow space separated from the farrowing pen, equipped with a heat lamp. Baby piglets can get in and out but sows can't.

to climb out. She paces up and down in her pen, pausing to rip the trough off the wall where you thought you had it securely wired. Or else she grabs her food pan and flips it to the far side of the shed. It is not a good idea to be in the pen with her at this stage. She hurts, she's feverish, and she may inadvertently bite you in passing. But she doesn't mean it* (see footnote).

When finally her pacing and chewing cease and her contractions begin, she will lie down quietly in her straw bed. That's when you climb into the pen with her, rub her tummy, and tell her what a fine pig she is. After awhile, a contraction occurs and, with a whoosh, out slips a little piglet. From the second it hits the deck, it's on its own unless you're there. The new baby immediately begins to struggle and stagger from mamma's back end towards that long row of faucets. Mamma conveniently draws her top leg up and forward so junior can crawl over the bottom leg to reach his nourishment. Very quickly, the piglet dries off and begins on his life of eating. You, of course, have helped him by toweling him dry, and clipping and iodining his navel cord.

For the next two to six hours, you will sit with mamma sow. In between birthing action, you read. Your thoughtful husband will bring you a thermos of coffee—sometimes spiked—and an egg sandwich on a paper plate. The egg will be runny, the fresh bread soft and crumbly and covered with a half inch of mayonnaise. However, you will feign gratefulness and eat the blobby thing, standing up in the farrowing pen, your hands scruffy and covered with iodine drizzle. At the first opportunity, try to toss some of the sandwich surreptitiously into the next pen. Eating eggy glop on soft bread is not a gourmet treat. Save the spiked coffee till last. If you are still in the pig parlor at regular mealtime, your dear husband will obligingly feed the human offspring and do up all the other chores—except for washing the dishes or the separator, of course.

Finally the sow is finished. Ten or more lovely babies vie frantically for room at the lunch counter. Eventually each piglet will settle for one teat and will always come back to that same one.

Make sure the sow has plenty of water, clean up her afterbirth (pitchfork it up and carry outside to manure pile), and spread fresh straw where needed. Deposit the little ones under the heat lamp in the creep a time or two. They'll usually crawl right back out to mamma, but from then on they'll return to the heat when cold.

*But it hurts.

Sometimes, the sow will have more babies than teats or a runt gets pushed off completely. Make sure the extra baby or the runt has swallowed a pretty good slug of that first colostrum milk from mamma. Then, as you leave the shed, pick up the runt and stick him inside your shirt. He'll ride there while you drive back to the house where you fix up a lidded box-house for him. Spread a large piece of plastic on the floor next to the heater stove. Put a towel-wrapped heat pad and some straw into the box. Cut a "door" in one end. Find another box, a tad larger than the house box. Cut the end out of the second box, line with newspaper. In the far corner, place a heavy low-sided dish of milk. The house box fits into the end of the milk box, giving a connected patio effect. The little pig, which by this time you've named "Portia," will snuggle down in her house and occasionally pop out to wee-wee on the newspaper (which you need to change often)* (see footnote) and drink milk from the dish. The first night it's best if you hand-feed her every couple of hours. Use a tablespoon and spoon in warmed cow's milk, or you can use a regular baby bottle if you wish. Some piglets take to a bottle instantly and others hate it. Within a day, the little piglet will be drinking nicely from her dish. Help her once or twice by stuffing her nose into the milk till she catches on.

A baby pig makes a fine temporary pet. She'll adapt to anything. She likes to be scratched on her tummy and loves to be rocked. Unlike a human offspring, a baby pig will wiggle and squeal to get down when she wants to go potty.

There was an old sow in the barn
Had twenty-three piglets by darn.
"I curse that old boar,"
Said she with a roar,
"Oh, when will ever I larn!"

*Newspapers, paper sacks and anything else of burnable paper is always saved for fire starter, bird-cage liner, and pig-box flooring. If you run short, visit the post office, where you'll find a trove of paper in the post office trash bins.

SECTION FOUR
Spring Farming and Crop Planting

Wherein one learns how to cope with mud, tired husbands, and emergency breakdowns of equipment. One also learns instant weather appraisal and how to "talk weather."

Spring farming and crop planting come at an inconvenient time of year—usually spring. Even if the calves are all on the ground, lambing is over, and the sows have all pigged, your husband is just not READY for planting. For several days now, the conversation has been about the weather. It is absolutely essential for the good Country Woman to learn to "talk weather."

In early spring, your man will cast an eye skyward, squint at the clouds, and sniff. Sniffing is more prevalent in spring than at other times of the year. As he sniffs, he opines that spring is on its way and that a storm is in the air. He tests the ground for moisture and worries about wet weather and scoury calves. He tests the ground for lack of moisture and worries about drought. He observes snow clouds and worries about a spring blizzard and his meager supply of remaining hay. Your job is to nod in sympathy, cluck in agreement, and keep him fed and fit if possible. Because at that certain pause between good and bad weather conditions, your man will hit the crop-planting trail.

Suddenly the ground has thawed and awaits the plow, the fertilizer, and the seed. Every year, the earth resists this intrusion. Mud puts out oozy tentacles that glom on to everything. For days, your man has fretted that it's too wet to plow. He tinkers with machinery—oiling and bolting and sharpening. One morning, the dawn comes up smiling. So does your husband. Happily, he gulps his four eggs, six pancakes, hash browns, juice, coffee, and a quarter pound of bacon. Rising from the table, toothpick tilted rakishly, he claps hat on head and strides out to meet the challenge of the tractor.

You watch as the tractor engine roars to life and away goes your knight on his tin horse* (see footnote). An hour later a totally demoralized husband returns to the house, spirit shattered. Drop everything instantly, pour coffee, and open the cookie jar. "What," you ask, "happened?"

Following a string of hair-raising expletives, you learn: a) the tractor quit cold, or b) the tractor has a flat tire, or c) the plow broke entirely in half, or

*All men are knights when driving a tractor.

d) (and most likely) the tractor is hub deep in gluey mud and your man has spent the last half hour alternately digging, cursing, and kicking the defenseless machine.

If the problem is that the machinery has somehow broken, you are elected to "go to town for parts." You will be told to get a No. 273 dizzel for a 1979 tractor. In the meantime, husband will pill scoured calves* (see footnote) and put out some salt. He says that's why you have to go to town—he's too busy.

In town, the local garage is staffed with men of amazing perceptive powers who will know exactly what a No. 273 dizzel is. The garage men are always patronizingly kind. Patiently, they listen to your requests knowing you will inevitably have the wrong information. This time, though, you have carefully extracted from your husband a complete description and model number of the part needed. Confidently you smile at the parts man and announce your request. Your smile freezes as you hear that the dizzel comes in right or left, with or without a toggle. You pale—just like in stories. You haven't the faintest idea whether it's a right or left. You try to phone your husband. Naturally, he's out pilling the calves. And he hates cell phones.

His last words were to hurry up—he's got to get that crop in before it rains again. Should you light out for home instantly? Or should you take four dizzels—two with and two without toggles? Desperately you conjure up a picture of the tractor and your husband. Just which side of the tractor anatomy did he pull that dizzel from? It's no use. You can't telepathize the information. You tell the parts man you'll take all four dizzels and bring back—right away—what doesn't fit.

As you are about to hop back in the pickup, you notice you've got a great wad of some kind of sticky black substance all over your front. You deduce it came from the home dizzel which you had carried warmly clutched to your person. The dark globs intermingled with the speckles of cherry juice (you had been constructing a cherry pie when the call for help came) make interesting, if tacky, designs. You feel, however, that these are proud emblems of the sacrifice you make for your beloved. That is, you feel that way till you are hailed by Mrs. Super Rancher. It's hard to hold on to your

*To pill a calf: First, catch animal. Then, grasp him by his front legs and upend him so he appears to be sitting on your shoes, his back pressed against your legs. Reach in shirt pocket for scour pill. Stuff pill down throat of calf. Try not to let him bite fingers. Meanwhile, keep a sharp eye out for an irate mamma cow. Last, let him go.

Usually you don't pill the calves, unless they are very tiny or very weak. Mostly your job is to follow along, shivering in the cold wind, and hand pills to husband as requested.

noble self-image when you turn to face her. She, of course, is clothed. You are barely dressed by comparison. Garbed in a reekingly elegant pants suit, she wants to know if you will bake some cookies for the church bazaar. Caught with no defense and totally intimidated by her warm, super friendliness and that gorgeous outfit, you agree to create four dozen cookies for Tuesday. Which is tomorrow.

You and your box of dizzels escape into your mud-splattered pickup. As you drive homeward, you have a semi-hysterical fit of giggling. Back at the ranch all is normal chaos. Your man is a picture of mud from futile efforts at extricating the tractor, which he stuck in the mud right after you left.

Humbly you offer the box of dizzels, while framing your explanation for bringing back four of them. He doesn't listen. Happily he stirs around and pulls out one of the dizzels—without toggle.

"That's it!" he says. Then he picks up the box and stirs again, brow furrowed. With a sinking heart, you timidly ask what it is he's searching for.

"The flange! The flange! Where's the flange?"

You knew it. All along you really knew it had been far too simple a trip. It now appears you must return to town to get a No. 273 dizzel without toggle, but WITH flange.

Keep a tight grip on your impulse to kill, and comply without even changing out of your cherry-dizzel-crease outfit.

At nine o'clock that night, your exhausted man tells you how good the pie is as he finishes off his second piece—with ice cream. Your day-long mad dissipates. Reaching for a toothpick, your knight-of-the-tractor leans back in his chair. You prepare to listen closely to the account of plowing, planting, and getting stuck and unstuck.

GOING TO TOWN FOR PARTS

Whenever the tractor quits or balks,
Or the mower refuses to start,
I'm the one my husband talks
Into going to town for parts.

No matter I'm buried clear up to my eyes
In bread dough and Pillsbury flour,
I dust off my hands, out the door I fly,
Vowing return in an hour.

While my knight of the tractor comforts his steed,
In the pickup I roar into town,
A Good Woman off on an errand of need
To the store where parts can be found.

A Part Man's a smart man who knows all factors
Of flange and bolt and U-joint
For healing the wounds of old, broken tractors,
He patiently waits as I point ...

"That left-handed flange with a gasket and hose,"
I say with a confident air.
With a withering glance down his hawk-beak nose,
He asks, "Just one or a pair?"

Confusion wells up and I know I'm pathetic,
"Oh, both," nonchalantly I say.
Twenty miles to drive home; my life is hectic,
I've lost the best of the day.

Back at the ranch, I seek out my man
"Look, one of each," I purr.
He takes the flange with an eager hand
Then utters a tasteless word.

"A flange is no good without some bolts,"
He growls with amazed disdain,
"Anyone could see, even fools and dolts."
My ego starts to wane.

Three trips to town and one flat tire,
And what do I find in the end?
The tractor gets fixed with pliers and wire …
No wonder I'm round the bend!

SECTION FIVE

Starting the Irrigation Water

Wherein one learns that water must be managed and controlled. The good Country Woman learns to help tear out beaver dams, pick rocks, and ignore black moods of the damp Country Man. Wherein one also learns about toting irrigation pipe and that a sprinkler system wants to hurt you.

Irrigation ditches: those peculiar, snaky-looking channels zigging through the hay meadow and circling the grain fields.

In spring, irrigation ditches on the ranch or farm must be cleaned and water introduced into them from the source: a river or creek. The main ditch, termed the "ragged-assed ditch" by the local male wits, serves a large number of farms and ranches. It travels for miles, beginning where the river tumbles from the foothills of the mountains. Each spring, the ditch must be cleaned of the amazing amount of rocks, trees, twigs, leaves, and miscellaneous debris accumulated over the winter. Cleaning the Ditch becomes a community effort for the men. You will also share in the activity, never fear.

Each farmer or rancher on The Ditch has a certain number of inches of water he's allowed to take out* (see footnote). Since irrigation water is the lifeblood of a ranch or farm, an occasional difference of opinion can develop over just who gets how much water and when. Those owning land located near the beginning of the main ditch tend to be calm, friendly men able to view any situation with an objective eye. As the ditch proceeds towards those ranches or farms lower down the ditch, tension builds among otherwise friendly neighbors. The last man on the ditch spends a lot of his time cursing, jerking out neighboring headgates, and insisting on his "rights"** (see footnote). The low man on the ditch is apt to develop serious personality problems. It's a lucky woman who lives on a place on the upper end of the main ditch. She's even more fortunate if he's a dryland rancher or farmer.

Before Ditch Day actually arrives, several of the neighboring ranchers "stop by" to see your husband. You are prepared with a full cookie jar and plenty of coffee. Seated around the kitchen table, the men sidle into an

*How those inches are measured remains a total mystery to all ranch and farm women.

**Shovel-dented skulls are not uncommon among irrigators.

agreement that TOMORROW IS THE DAY. The decision is reached during the last five minutes of the conference, which has lasted for two hours.

As a loving helpmate, you do not hurry the confabbing along. However, a private prayer is allowed, urging HIM to see fit to end the conference before 11:45 a.m. At that time, if the kitchen is still full of persons, country custom says you must invite ALL of them to stay for dinner. The conferees will look up, focusing upon you personally for the first time that morning. All will politely decline your invitation. But the first "no" doesn't count. You must eagerly repeat the invitation and elaborate by mentioning some part of the menu such as: "There's chocolate pie for dessert!" Again the guests will decline regretfully but with nostalgic comments about chocolate pie and congratulations to your husband (not to you) about what a good woman he has.

After the second refusal, it's reasonably safe to assume they'll all get up and leave shortly. The men initiate the leave-taking ritual by standing up, placing hats on heads and carrying used cups to the sink (sometimes). They then bunch up by the kitchen door where the final arrangements for next day's program will be completed. It's quite possible your hospitable husband may still persuade one or two men to "stay for dinner"* (see footnote).

Early the next morning a gaggle of men begin ditch cleaning. With backhoe and shovels, they will scoop out a wide assortment of debris that has accumulated over the winter. Ditch banks must be shored up and culverts cleaned.

As he leaves, your dear husband suggests that you start cleaning the lawn and garden ditches "if you've got time." Long ago you somehow learned to avoid discussion of "time" and if you've got it or not. Therefore, after noon dinner, leave all the housework and go get ready to clean ditches.

Your ditch-cleaning and irrigating costume are the same. Thigh-high, heavy-rubber, ugly gray boots will chew your feet to a blistery pulp in no time. For some reason, it's not possible to buy boots other than too big. At each step, your foot comes up followed a half second later by the boot itself, making a shuffle-smack-thunk sound as you walk. Wear stout leather gloves, which will get soaked, but will help prevent more serious wounds from sharp sticks and heavy rocks. Caution: Do not use rubber gloves. They're slippery and you tend to drop hurtful stones on your toes.

*Always be prepared for extra mouths with a backlog of eggs, bacon or sausage, and instant pudding.

In your jacket pocket, carry some chocolate bars* (see footnote). At a casual angle over one shoulder, wear an irrigating shovel** (see footnote). Over the other, a manure fork is balanced. Don't forget your hat—anything with a brim and that doesn't fall off easily. Your pretty, red-and-white speckled, straw garden hat works well because it can be tied under the chin by its chiffon ribbon. You look a little like a chubby, bent-elbowed, rubber-booted coolie.

Of course you know why you have been elected to clean the garden and lawn ditches. Its source of water is diverted from Antelope Creek rather than from the Main Ditch. And it's ALWAYS full of water so when you clean, you wade. This ditch wanders through the cottonwoods and meanders among the apple trees, across the nearest meadow, and thence to your garden and lawn. Jump in at the garden end and begin scooping out sticks, twigs, leaves, and rocks. The rocks tend to become heavier and heavier as you go. An hour and a half is the outside limit for stooping and lifting. So sit down for a chocolate-bar break. As you try to get up to return to work, you note you've acquired a semi-permanent forward-leaning posture. Still, maintain a cheerful attitude about it all. It's a lovely day. Except for the pain in your back, it's great to be outdoors.

You're nearly through—just the half acre through the cottonwoods to go. You talk yourself into a rather sickening self-congratulatory Pollyanna attitude when you run into a beaver dam. The dam consists of a solid mass of interwoven sticks, twigs, leaves, and small logs, which have got to come out. A beaver dam is several feet long and wide and thick and you must unthread all those interlaced sticks, twigs, logs, and slimy moss like a game of jackstraws. Always you manage to yank on a key twig, pulling a hunk of the structure loose with a whoosh that sloshes icy water over the top of your boots. Your cold clammy feet now weigh a ton each. Go right ahead with your choice of curses and screams. No one will hear you and the extra emotion will help keep you warm.

With commendable bravery, you keep at it. Finally the ditch is clean, except for that one small boulder you simply couldn't budge. It's past chore time as you slish-slosh home to find your loving husband has done the milking, the separating, and shut up the chickens. All you have to do is fix supper.

*Chocolate bars are a must on most excursions. They can be used as an excuse to "take five" when you've reached the limit of endurance and will fend off convulsions from hunger pangs.

**Irrigating shovel: A remarkable tool, sharpened on its business edge. It's used to shovel, chop, scoop, pound, mash, lean on, and kill the occasional rattlesnake.

When your man announces he's been back for ages and what kept you so long, do nothing rash. Calmly and wordlessly remove your iron-rubber boots and wet socks. Pad into the bedroom and slip into a pair of HIS big, woolly socks and HIS fleece-lined slippers. Then just as calmly, march on grateful feet back into the kitchen, reach into the lower left-hand cupboard, and extract the large bottle of medicinal bourbon. Measure two generous toddies, sit down with your man, and prepare to exchange ditch stories. Supper can wait.

The big irrigator went *wham*!
His shovel collapsed on the dam.
He jumped in the ditch
"You son of a b . . . h,"
He cried; now the beaver is Spam.

SECTION SIX
Hired Hands

Wherein one learns that the vagaries and varieties so richly dispersed among humans abound ten-fold in hired men and/or couples.

Spring brings variable weather, new calves, lambs, piglets—new beginnings of all kinds. For many ranches and farms, spring also brings hired men. If your operation is a one man, one wife, one family outfit, your husband likely needs extra help only in spring and summer. As YOU are always the instantly available hired hand, you think you will be glad to see a hired man come on the place. You've forgotten what happened last year.

Calving, lambing, pigging, cleaning the ditches, milking the cow, getting the wood up, barn cleaning, irrigation, spring planting and farming, and fencing and haying are a few of the chores another hand could help with. Both you and your man are developing upset stomachs and short tempers trying to do all the additional work, attend all the spring functions at the school on behalf of your two sons, and keep up with the usual time-consuming chores.

So one day your husband goes to town to "hire a man." The country hiring process effectively eliminates middle man services of an employment office. Your husband merely drops in at the Sheepherder's Tavern, the Stockman's Bar, the Timber Inn, or the Longhorn Saloon. In each establishment he purchases a drink, hooks a foot over the rail, leans an elbow on the bar, and commences discussing the weather, calves, and spring planting with Sam the Bartender. Eventually the subject of hiring a hand slips into the conversation. Casually your man will ask Sam if he knows of anyone looking for a job* (see footnote).

Often Sam is able to produce a man for hire on the spot. It happens that at the other end of the bar, there leans a cowboy. He is drinking a short beer—slowly, striving for nonchalance. "Here's a fella just said he was lookin'," says Sam.

The job interview is then conducted on the spot, but first your husband orders another round of drinks. (The cowboy/farm hand changes his order from beer to hard liquor.) As possible employer and potential employee negotiate, Sam and any other loose men standing around interject helpful comments about the long hours, low pay, and horribly hard work to be endured if one is foolish enough to hire out to your husband.

*All bartenders in male hangouts in country towns have been there forever and know everything.

If the first saloon stop does not yield a hired hand, your dauntless mate will try the next, and the next, and the next, if necessary. Rarely can you expect him home before evening chores. The later his arrival, the cheerier his outlook—with or without a hired man.

The thinking Country Woman will have a thick stew and hot coffee ready the second he walks or lurches in. If your husband has had a successful trip, he will be followed through the door by the latest "hired hand."

Hired hands can be solid, faithful, hardworking, salt-of-the-earth men who will devotedly care for the livestock and the land. However, too often a man is hired who demonstrates a definite personality quirk within the first day or two. His quirk usually interferes with his work.

As a Country Woman, you learn to categorize the transient help into types. There's the kind you think of as the "Slobberin' Joe" type. Slobberin' Joe never works anywhere longer than a month and a half. He derives his descriptive appellation from the fact that his constant chatter causes him to drool around his loose false teeth. He loves best to talk while chomping away on a giant stalk of celery, teeth clacking and moisture oozing from the corners of his mouth. Joe is also a butter-licker. After stabbing a hunk of butter from the dish, Slobberin' Joe juicily licks his knife and then stabs into the butter again for another gob.

If the new hand is not a Slobberin' Joe type, he may qualify as an "If Only" sort. He is a man able and willing to tackle anything—he says. A heck of an irrigator—he says, "if only you had a sprinkler system." Often, he will relate at length how good he is with horses—"if only" he hadn't hurt his back when his last employer made him "lift those heavy bales." His tale of personal prowess goes on and on.

You, as a good Country Wife, are privileged to listen to If Only at the dinner table because naturally you get to cook for him as well. The If Only type is especially pleasant during meals. He considers it only courteous to include you in his endless commentaries, which means he compares anything you cook with the way his previous employer's wife (or cook) prepared the dish.

Fortunately, just as you are about to feed the man rhubarb-leaf salad, he quits. But with a good excuse, of course. His back hurts, his teeth need relining, he needs a haircut, his sinuses are acting up, and his dear old mother has passed on. He draws two months pay from your over-generous husband for his month and a half on the place, packs his gear, and waits underfoot in your kitchen. While he waits for you to finish the dishes so you can drive him to town, he comments on your dishwashing style.

After If Only's departure, your husband hires another man who turns out to be a "Watcher." A Watcher, married or single, is usually quiet, soft-spoken, and terribly agreeable. His very quietness gives a misleading impression he can and will actually work. You will never know, since it's a rare moment that he turns his hand to any task if you or your husband are around to do it instead. On the day the Watcher is sent to the back pasture to get fence posts from the pile there, your mate has to go search for the fellow an hour later. The Watcher can't find the back pasture. Leading the way, your husband personally shows the Watcher the pasture and the posts. He (your husband) then loads them on the pickup while the Watcher watches. Occasionally the Watcher stoops to help lift the back end of a particularly long and awkward post. Or perhaps he'll let down the end gate of the vehicle to make it easier for your husband (or you) to slide the poles onto the pickup bed. The Watcher is semi-alert to any little thing he can see to do to make it easier for someone else to accomplish a chore. Often the Watcher combines watching with a persistent tuneless whistle.

Even though the Watcher watches such a lot, he suffers from dim vision and hearing loss, likely due to the constant whistling. Sent to check the ewes, the Watcher wanders right past a sheep bleating like mad and up to her neck in the slough. The Watcher reports all is well, but you can hear the commotion from inside the house. Respond to your uneasy feeling and go check the pasture yourself. Likely you will wind up alongside the ewe in the slough. Heave and shove her out, extricate yourself, and take your soggy, gorpy self back to the house. As you drip up the path, the Watcher will hold open the backyard gate for your passage.

"Looks like you've been in the mud," he'll observe.

The Watcher is always prompt for meals. From a slow, tender step that looks as though he's suffering from ring-bone, he swings into a smart canter when the dinner bell rings.

Very soon, you again haul the "hired man" to town, because your husband has fired him.

Next to follow the Watcher may be "The Decorators." Decorators always come in pairs. The man works on the ranch. The woman works anybody she can. Her aim is to extract as much as possible from the ranch owner. Always she begins her residency by redecorating the bunk house.

You, personally, have taken pains to keep the bunk house cottage freshly painted and repaired. Yellow curtains with white daisies blend with pine woodwork and cabinets in the kitchen. A stone fireplace in the living room complements the open beam ceiling. It's a cozy nest, furnished completely. But a Decorator is never satisfied. Usually she favors lavender and orangey-pink, which she proceeds to calcimine all over the knotty-pine living room walls. Naturally she charges the paint to your account at the lumberyard.

After the Decorators leave, it requires considerable expenditure and time to have the woodwork and paneling electrically sanded and refinished.

Then there are the times your husband hires a wandering sheepherder usually named Sven or Olav or Sig. This type of individual can drive you, as an otherwise reasonably sane Country Woman, into a nervous collapse. Sheepherder Sven is often mentally peculiar and downright crazy if he drinks. Sven will talk to himself at length, which doesn't bother you too much since you tend to do that yourself on occasion. But Sven argues with the voices in his head in his native language.

If Sven manages to imbibe too much liquor (his favorite pastime), he tends to become delusional and proceeds to disrobe entirely. Generally, that sort of display takes place in the town park where he then leaps faunlike among the pines and cottonwoods. The town park is located across the street from the town grade school.

But when sober, Sven is a good hand with sheep. It is up to you and your husband to keep him pure, which means checking his clothing and his ancient pickup at frequent intervals for contraband booze. Eventually, the ranch loses Sven's services due to his untimely removal by the authorities to the local funny farm. One weekend, Sven went to town, got drunk, and appeared stark naked at the coffee shop in the town hotel and a lot of people complained. Early morning streaking by a grizzled old sheepherder apparently depressed other patrons' appetites.

Sooner or later in the round of hired hands, Grammaw Charlie arrives. Grammaw Charlie paints. That's his one talent. He loves to paint. Hired to help with fencing and haying, his back goes out the second day on the place. Your husband puts him to painting the barn.

"Grammaw" earned his nickname due to his fussing habit. He's a champion fusser. He fusses about a spot of mud on the kitchen floor. He fusses about a mop until you fetch one for him. Then, fussing, he scrubs up the mud. He fusses because the mop is tattered and then he fusses because you didn't put down newspaper on his fresh cleaned floor. Charlie fusses with each stroke of paint applied to the barn. He fusses about the brushes, the color, the ladder, the paint thinner, and the hazards of inclement weather, such as a cloud, on his artwork. Your husband gets rid of Charlie by recommending him to a rancher at the other end of the county as a "good man with a paint brush."

Each hiring season, your ranch or farm acquires a Tex, or a Pecos, or a Slim, or some other colorful character. These men are always expert stockmen, they claim. Last Chance Jones claimed he was the world's greatest hand with livestock. He favored a sparkly hat band on his huge black cowboy hat that complemented a black satin shirt with rhinestone buttons. Last Chance was a great rope twirler and fascinated both sons for two days. On the third day, however, the Sheriff drove into the yard and collected Last Chance. It seems the world's greatest roper and stockman was wanted for check artistry.

Odd personalities abound among transient hired hands, but since yours is a small outfit, you don't have near the problem your neighbor has. She contends with a whole wad of extra men and women all summer, especially a series of cooks.

It's usually women who hire on as cooks. A good cook can be forgiven almost anything on a big outfit. Your neighboring Country Wife felt she'd tolerated enough when she found out that Cookie charged the men for mid-morning coffee and doughnuts. Your neighbor was puzzled as to why the men meekly paid up until she discovered the reason at three a.m. one morning. It seems Cookie dispensed nighttime favors to the men on a come as you are basis.

For some reason, some women ranch cooks often have a history of selling personal favors commercially. Usually cheerful and easygoing, such a lady will often advise you on her former occupation. Once, while visiting your neighbor, her cook fascinated both of you with a detailed account of how to procure, what to charge, how to dress, and what to say if questioned by nosey police. You and your neighbor took it all in. As a good Country Woman you should be open to educational experiences of all kinds.

Sometimes the cook is an excellent chef but prone to drunkenness and fits of temperament. The least upset causes Cookie to reach for the bottle, whereupon, you, as instant substitute, have to fill in.

A drinking hired hand is unendingly exasperating. Often capable and alert (when sober), he quickly becomes useful and dependable on the ranch or farm. You notice that the pressure of constantly running to keep up with the chores is beginning to ease. You are surprised when a whole month has gone by with no serious upsets or irritations. Then one day, right after payday, the hired man asks for a day off in the middle of the week. (It's also the middle of haying or some other major project.) He desperately needs, he says, a package of cigarettes, a haircut, a postage stamp, some new underwear, or a watchband. You feel an instant warning flutter in the pit of your stomach. You offer to bring him cigarettes and mention that you're a whiz at hair cutting. No matter what you say, he has a flanking argument, including the passing of his dear old mother.

He goes cheerfully off to town promising to be back in time for supper. Three days or more later, the phone will ring. It's Sam the Bartender inquiring if you want the man back or should he call the sheriff. Your husband may take him back hoping to get another one-half month's work out of him before the next binge. Usually, it's your spouse who goes to town to collect the inebriated hand. But sometimes, you, as the person-who-can-always-be-counted-on, find yourself driving to town after a sick and broke cowboy. You've had experience before, so you know what to expect. Expectations are fulfilled. Sam, the bartender, has stashed Willie, the hired hand, in the "snake pit"* (see footnote). Sam helps you lead Willie to the pickup. Meanwhile Willie sings his version of "On Top of Old Smokey."

It's a warm day and Willie has neglected bathing recently, so be sure to crank the windows down. Sam straps him onto the seat with the seat belt. Between verses of his song, Willie pleads for "one more l'il drink." You say absolutely nothing. Think of Willie as a difficult critter, which has to be corralled and trucked home. Never try to talk, reason, or plead with a drunk. Repeat: Do NOT respond, no matter what he says.

Back the pickup away from the curb and drive around to the United States Post Office. Like a dope, you forgot to go to the post office first and pick up that package of vaccines. But you must chance it. Willie is strapped in, his song reduced to a dull rasp. Quickly hop out of pickup, scuttle into

*Snake Pit: A small dark room with benches where drunks sleep it off or climb the walls, according to their needs. Kindly, Sam brings them coffee and soup till they're fit to stagger forth.

post office, and pick up package. You're breaking for the door when you are accosted by Mrs. Super Rancher. She wants to know if you're coming to the Club meeting tomorrow. Chat superficially, but try to hurry her along. You're terrified that Willie will escape and you'll have to give chase. You and Mrs. Super Rancher reach the sidewalk in front of the post office where your outfit is parked at an angle. At that moment, Willie leans out the open window and barfs.

Feeling pretty nauseous yourself, turn to Mrs. Super Rancher and remark lightly, "See you at Club." Quickly dart around to the driver's side, pile in and back away from the curb. As you maneuver the vehicle into the driving lane, you catch Mrs. Super Rancher's eye. Flash a super-charming smile, twinkle your fingers at her, and gun the motor.

No matter how odd or peculiar the men may be, as a good Country Wife you must be prepared at all times to act as den mother to hired hands. You may "boss" a hired hand if you stick to orders regarding meal times or the cleaning of muddy boots. You can go so far as to "request" a load of wood to be carried in for the cook stove. But never, never order a hand to do any of the "real" ranch work* (see footnote). Such directives are the prerogative of the ranch or farm man only. It is sometimes OK to relay a message from your husband to the hired hand in an *extreme* emergency such as the house burning down, the basement flooding, or if your water just broke.

If you, as a female, attempt to direct a hand to start fencing the lower pasture, for example, you will find the fellow has developed acute deafness. Recommendation: If you want the fencing (or any other task) started and your man isn't home to issue the order, there are alternate ways to approach the problem. First, obviously park the pickup in front of the shop door. Then fetch and carry fencing tools and wire to the vehicle. Load some poles and, lastly, throw in a sharpened spade. You have arranged it so the hired man notes all this activity. Mention to him you're going to the lower pasture to do some fencing and laughingly joke you hope you don't cut your hand open on the barbed wire "like last time." Give the hired man plenty of time to ask "if you'd like some help." To which you reply, "Do you have time? I'd appreciate it. If you'll drive, I'll open the gates." Whereupon both of you commence to the lower pasture where you hold and hand things while the hired man stretches wire and pounds staples. When you become bored, mention you have to check on the bread rising and then walk home. If it's

*The only time a woman may give orders is if her husband is sick, in the hospital, or dead.

too far to walk, take the pickup and lie that you'll be right back. At home, wait for Number One son to arrive from school and send him after hired man. Number One son is twelve and loves an excuse to drive anywhere.

SECTION SEVEN
Planting the Garden

Wherein one works like a fiend after coaxing husband (diplomatically) to plow. One plants rows and rows of potential foodstuff and resolves NEVER to let one tiny, miniscule weed mar the perfection.

Considering all the peculiarly titled country-western songs you hear disturbing the peace, there should be one labeled: "When It's Garden Planting Time, Smile and Let the Dishes Go Undone Blues."

This year, the garden shall be beautiful—your annual noble resolve. A garden is a major part of a Country Woman's spring and summer life. Lucky and blessed is the woman with a green thumb. For her, everything grows, the birds sing and eat only the slugs, and weeds are timid, fragile growths of no importance. The neighbor's garden is one of those—a magazine picture of planned beauty and productiveness. Your own garden has a more utilitarian appearance. But this year you determine things will be different.

Stashed away in the cupboard behind the ice-cream freezer on the porch are $76.50 worth of seeds ordered from five different seed companies. The circus-bright packets hold exciting varieties of vegetables, cucumbers, cantaloupe, squash, four kinds of corn, pumpkins, peppers, and spices and herbs of all sorts. Plant these practical seeds in a large sunny plot beside the house. Along the walks and in the front and back yard, go wild with a rich profusion of flowers and shrubs—all planned to bloom with glorious color all summer long.

This year you are determined to be efficient. Commence by drawing a Master Plan on paper. After the Master Plan, the next major hurdle is to prevail upon the Master of the House to implement your Master Plan by plowing the garden. For some reason, garden plowing is always last in your husband's mind* (see footnote). First he must plow the oat field. Then a spare part must be purchased before the tractor can function again. Or he claims the ground is too wet, or there was a frost and the ground is too hard. Or the widow woman down the road has to have HER garden plowed. "She's alone," your man says. "It won't hurt to help her out"** (see footnote). If gentle

*Investigation by a select committee of Country Women reveals that resistance to garden plowing stems from childhood. As boy children, the men were forced to weed, till, hoe, plant, and water. Pay no attention to this, however. Force your sons as well.

**You can't beat this particular excuse, so send along a jar of last year's mint jelly and a loaf of fresh homemade bread.

pleading at breakfast does not produce results, the thinking Country Woman must resort to other tactics. Have a headache when you retire because of "worry about putting the garden in." This works sometimes, especially if you've thoughtfully purchased a copy of *Playboy* and left it on your spouse's pillow. Next morning, bake an enormous chocolate fudge cake and smear on thick, chocolate fudge frosting. Mix up some pink icing, decorate the cake edges with pink roses and in the middle, write:

Grand Prize
Garden Plowing

Do not serve till after plowing is actually finished.

After all the hard won plowing and fertilizing, the garden is left with big black curds, which have to be tilled to manageable size. Then the soil must be raked level and smooth. Dig out your rubber gardening boots* (see footnote), which will be found under the stairs leading to the basement. Find some thinish leather or plastic garden gloves and don your striped bib overalls. The red-speckled straw hat with the chiffon tie completes the ensemble. Stock your pockets with tissue because all the bending and stooping makes your nose run.

A runny nose is the least of the bodily disorders that occur as a result of gardening. When your back reaches an unbearable level of pain, be ready with certain remedies. First remedy: Try not to bend and stoop. Drop to hands and knees and crawl. The rows may become crooked, but pay no attention. Remedy Two: At 4 p.m., after a brutal gardening session, go to house, draw a deep hot bath to immerse self in for thirty full minutes. Do not start timing till you are in tub. Take with you some reading material that won't tax your mind too severely such as a mystery, or a gothic novel, or a favorite romance tome. Remedy Three: When it becomes apparent you are crippled beyond a home remedy's capacity to cure, go see the chiropractor. He will push, pull, and wiggle your warped spine back into alignment. In addition, your disability will then achieve the status of needing medical attention.

Direct orders accompanied by awful threats encourage both sons to help with the back-breaking labor. Finally the lovely, thick, black-velvet carpet of loamy earth smiles up at you. Whip out the Master Plan and the seeds. If you are one of the fortunate green-thumb Country Women, you know exactly what to do. Automatically, the spacing, the depth, the loaminess

*Garden rubbers: Either those old red rubber boots you wore years ago in high school or cut-off irrigating boots that developed leaks, making them unfit for ditch work.

or non-loaminess of the soil, and the proper sequence of the planting fall into place. You, however, consult the directions on the backs of the seed packets. Rake, dredge, poke, hoe, pat, and tromp. After awhile you will have impregnated the black earth with thirty-four semi-straight rows of seeds. The garden fairly hums with life.

Faithful garden watering is a must. Assign watering chores to the offspring, which helps teach them the lure and lore of growing things.

Put in heaps of potatoes, onions, corn, and green vegetables. Hardy seeds can be planted outdoors early in the spring. Those seeds started in the house require a different approach. Gather peat pots, paper cartons, Styrofoam cups, and odd-shaped plastic cheese dip, sour cream, and yogurt containers. In these, plant herbs, spices, flowers, and especially tomatoes. Such plants must be started in the house in order to reach enough maturity to be transplanted to the outdoors.

Start the tomato seeds in individual earth-filled peat pots and set the pots (at least twenty) in your extra-large baking pan or in old refrigerator trays. Since all the southwest windows and ledges are overflowing with potted things, the tomato-plant trays live with you in the kitchen. One tray filled with little thready green plants decorates the kitchen table edge next to the window. Eating is still possible, if crowded. Keep a vigilant eye on Blossom the housecat. It is best to leave a special eight-inch space in the sunniest window for Blossom to sit. Otherwise she deliberately paws to the floor each tender potted plant. Firmly discourage her from eating the fresh green shoots. Smacking her with a fly swatter is usually effective, but sometimes you can't find the swatter. Keep a spray bottle of water handy. She hates being sprayed. Merely reaching for the bottle will make her throw up her paws in disgust and stalk off.

Successful tomato raising is matter of pride to the Country Woman. Once transplanted in the ground, protect the little dears from high winds and possible unexpected late frost. Stick some of those old leftover shingles around each plant at all four points of the compass. If you're out of shingles, look out behind the old well house and swipe some of the worn-out tires your husband has been saving. These make excellent protective wind breaks. Stack two high around each plant. Till all possible danger of frost is past, each and every night cover the tomato plants with old sheets, gunny sacks, old dam canvass, or old tablecloths. Each and every morning remove them. Almost always you forget to cover the plants till you're ready for bed and can't find your slippers. With true Country Woman resourcefulness, slip into your husband's footgear and his big, warm bathrobe. Find the flashlight, scuff into the garden, and flop the coverings over the plants. Ignore the fact that the hem of your husband's robe is becoming soggy and think only what a truly conscientious person you are. Remember, frozen tomato plants are one of the worst tragedies that can occur in a gardening career.

As a true-blue gardening Country Woman or merely one with good intentions, you must construct a compost heap. In a corner of the garden reserve an area about two kitchen tables in size. Gather grass clippings, old bedding straw from the pig shed, all the leaves raked from the yard in the fall, vegetable peelings, and some of the free sawdust you got from the sawmill. Pile all in alternating layers with manure . . . sort of like a garden lasagna. Any kind of manure will do, although chicken and sheep are considered to have the greatest fertilizing power. They certainly smell that way. Make the manure layers thin, and the leaves and other material six- to eight-inches thick. Some scientific super gardeners will put their compost in a giant box with hinged lid. When all inside deteriorates and rots through and through, super gardeners spoon out the stuff in measured amounts onto the soil. Pay no attention to all that. Remember the Country Woman's simple directive: Pile it up, let it rot, put it on. However, you need not bandy your methods to anyone. When the compost heap enters the conversation, always sound completely knowledgeable by discussing in positive tones, each layer of rotten stuff, its depth, its consistency, and its potency.

EXTRA DELICIOUS AND FATTENING CHOCOLATE CAKE TREAT

Whip a batch of cream. Add sugar and vanilla to taste. Serve over cake squares with shredded chocolate or chopped nuts or both sprinkled over. Top with a cherry. You'll think you've died and gone to heaven. (Good for soliciting garden plowing and varmint-catching, as well as other sticky challenges where Country Woman Diplomacy is needed.)

MILDRED NORTON'S RICH, DARK, CHOCOLATY, CHOCOLATE CAKE

1/2 cup shortening
2 cups sugar
2 eggs
1 teaspoon vanilla
1 cup BOILING water
1 teaspoon baking soda dissolved into 1/2 cup sour milk (Sweet milk may be soured with a tablespoon of vinegar, or you may use buttermilk.)
3/4 teaspoon salt
1/2 teaspoon baking powder
heaping half cup cocoa
2 cups flour

Mix shortening and sugar; add eggs and vanilla and beat. Add sour milk (with the soda dissolved in the milk first!). Sift together flour, salt, baking powder, and cocoa and add to the above. Beat well. LAST: Add 1 cup boiling water.

Makes very thin batter. Do not panic. It's supposed to be thin. Pour into a large, flat pan (approximately 9 x 13). Bake at 325 degrees, about an hour or so.

FROSTING
Melt 1/4 pound of butter. Add about a pound of powdered sugar and 1 1/2 teaspoons vanilla to the melted butter and beat. Add canned milk or cream and beat like mad. Add only enough milk or cream to make the frosting spreadable, not sauce-able.

SECTION EIGHT
Spreading and Harrowing

Wherein one learns that operating the manure spreader and harrowing the cow pies are two country chores the Country Man ACTUALLY believes you're smart enough to do.

Ranching and farming are occupations constantly concerned with animal manure. Its collection and dispersal are important springtime duties. "Money-in-the-bank" is an oft-repeated slogan. This natural fertilizer will make the grass green and the alfalfa grow, and you can sell any overabundance to city gardeners by calling it "steer manure."

Throughout the winter, from the cow barn, the pig shed, the chicken house, and the sheep shed, the stuff has been regularly scooped up and tossed onto an ever-growing pile. The cow barn even has a special window out of which to pitch manure. By spring, the heap reaches as high as the windowsill. The roosters enjoy climbing to the peak of manure for a good crow and pigs will scale the summit to jump through the open window into the barn to steal grain. The piles down by the pig sheds and the sheep shed have also reached impressive dimensions. All winter you've been wheelbarrowing the stuff onto the heaps. It is now time to utilize all this home manufactured fertilizer.

To the tune of terrifically clever remarks, your husband and Number One son fill the manure spreader one shovelful at a time. If you're lucky enough to have a front loader for the tractor, the job can be accomplished in no time. Number One son will miss a meal just to get to scoop manure.

A manure spreader is a long, usually green, box affair on wheels. Situated at the back end of the thing is a large, spiked, horizontal cylinder thing that looks like a huge bristled hair curler. Somehow this whole contraption turns when the spreader is hitched to the tractor and certain gears are engaged.

Try to arrange it so your husband will "let" you drive the manure spreader even if you have to hitch up the loaded spreader yourself* (see footnote).

Begin at the lower corner of the hay meadow and line out the tractor with the manure spreader following faithfully behind. Reach down, pull

*As a good Country Woman, you learn to hitch up and operate all the farm equipment. WARNING: Take Care! All ranch and farm machinery is stealthily hostile to women and will pinch, bruise, and bash.

that short-handled knob to half position, and commence spreading. Which means that as the hair curler turns on a steel track arrangement, the heaped manure is combed off the end of the spreader. The manure doesn't merely drop quietly to the ground, though. With joyful abandon, small flakes and globules of the stuff fly up and up into the air to rain back to earth. Most of the material flies straight up and you've driven ahead before it has a chance to fall back on you. But there are always occasional flying curds that seek you out, especially if you've just washed your hair* (see footnote). A wide-brimmed straw hat, leather driving gloves, overalls, and denim jacket is the ensemble of choice for this operation. Somehow the combination of fresh sunny air, the buzz of the tractor, and the excitement of the flying fertilizer arouses your urge to sing. Hymns and spirituals lend themselves particularly well, so as you sweep around the meadow, let go with a hearty "R-o-c-k-of-A-a-a-ges-s-s. . . ."

Through the winter, the cattle have been pastured and fed on some of the hay meadows. Cattle are roving creatures and have done a good job depositing their droppings at random all over the fields. Before the spring thaw, feeding the cattle demands skill, a sturdy vehicle, and a sound body. Both body and vehicle must withstand the spine cracking, teeth rattling, jarring, and thumping, which occur as you bounce over terrain littered with frozen cow pies. In winter, a fresh dropped cowpie turns to instant cement, which miraculously softens as the weather warms.

The problem now becomes one of dispersion. A cow pie that just sits there in its own juices eventually dries to a pancake, which smothers the grass under it. So one fine, warm spring day your man will ask, "Do you have time to harrow?"** (see footnote).

Harrows come in a variety of styles. A favorite instrument is that contraption of steel, grill-like arrangements alternating with hard-wood poles. The whole looks like a jail house's barred windows laid flat on the ground and linked together by a system of chains*** (see footnote).

Garb yourself in denim jacket, grab your straw hat, and wear leather gloves. No special requirements for coveralls or overalls. Regular jeans and natural cushioning will do. Mount the tractor steed and charge the field.

*If the wind is blowing, reconsider spreading for that day.

**As a good Country Woman, you always have time for a harrowing experience.

***Harrows can be home-built out of random pieces of iron pipe, wagon spokes, old tires, and old hot-water radiators. Country Men have been known to spend whole days happily knitting all this together.

(Your thoughtful husband has hitched the harrow to the tractor for you.) Begin at the fence corner and drive over the field in ever diminishing circles. The harrow bounces along behind, crashing into the cow piles, mashing and mauling them into acceptable sprinkles of manure. While harrowing, the Country Woman is completely free of the house for a time. As you watch the harrow chew away at the cow pies, the sun warms your back and a beautiful, peaceful chunk of time happens to you. Sometimes you think you'd like to write a book—if you ever get time.

Though little in life is for sure
Three things on the farm will endure:
Sticky mud to your thighs,
Ugly bugs every size,
And a steady supply of manure.

The farm-woman who's out on the tractor
Is happy except for one factor
When every last bump
Meets up with her rump
She's certain that someone has smacked her.

SECTION NINE

Instant Hired Hand

Wherein one learns how to become an instant expert at all country chores. One learns to accept second place to any country critters having four legs and a tail.

The Country Husband sometimes needs his woman to drive the tractor while he feeds the cattle, or vaccinate the calves while he holds 'em, or hold 'em while he castrates 'em and brands 'em. In any case, you must be an instant expert in all phases. You will learn to employ muscles not normally used by women or even elephants. If you're really good, you may earn that best of all songs of praise, "She's a helluva hand!"

Try very hard not to take personally the adjectives coloring the air, especially when working with the stock. A poor, innocent cow, sheep, or pig takes on amazing qualities and spurious ancestry when husband is discussing them in the corrals and through the chutes.

Of equal importance are the following commandments for an I.H.H.* (see footnote) woman:

1. Keep mouth shut and body alert and poised to jump in all directions.

2. Be brave, selectively deaf, and keep mouth shut when the language becomes more and more virulent.

3. NEVER try to lighten the situation with pleasantries or observations on the beauty of the day.

4. NEVER, NEVER point out an error, or goof, or alternate method of approaching the task.

5. NEVER, NEVER, NEVER giggle at the cute antics of the animals while your man is physically engaged in working with them. (Unless he laughs first.)

Even when observing the commandments, some days are unbearable. If, in spite of every precaution taken, the situation begins to deteriorate into an emotional catastrophe, there is only one thing left to do. Put down whatever implement you're holding and begin quietly to cry. Turn on your heel and go to the house. This action MUST be carried out without speaking a word and without looking back. Earplugs, at this point, are of particular value.

Your duties as an Instant Hired Hand occur on an emergency basis always. Your man saddles up to bring in the cows from the South 40. As he rides past the kitchen window, he bellows for you to come watch the gates.

*I.H.H.: Instant Hired Hand

Scrape the bread dough off your hands, grab your wide-brimmed hat, and don't forget gloves. (Gates have a tendency to impale you with splinters.) Gallop on foot for the corrals.

Across the lower meadows, you observe a thundering herd of cows and calves blasting towards the big pole corral gate. That's the very same gate you are prepared to swing open, then duck out of the way so you don't spook the critters. With flourish* (see footnote), you shoulder open the heavy pole gate and wrestle it to wide-open position. All goes well. The cows rip into the corral and begin to mill around. You slip to the gate to close it. (All gates are harder to close than to open.) Try not to panic as you keep an eye peeled for that mean, wild-eyed roan. With the gate halfway shut, the roan roars out an awful bellow and gallops towards the slowly closing aperture. As you stare in horror at the charging cow, you're also aware that your man and the REGULAR hired hand are riding up fast. As a good Country Woman, you will die before you let that darned cow escape through that gate! From depths you are surprised you possess, emit a horrifying howl. Leap two feet up and three over to land in front of the charging critter. Sweep your hat off your head and wham it in wide smashing arcs before you. If you can stand it, keep your eyes open so you can tell if your actions are effective. At this point, you will discover brand new Anglo-Saxon names for Bossie. But you win. The old cow circles away and back to the herd. Proudly finish closing the gate and assume your best air-of-nonchalance** (see footnote). Maintain your air as your grinning husband swings off his horse to observe, "That old slut is wilder than a peach orchard boar!" Take it for granted he means the cow.

Probably the least exciting I.H.H. duty is the thing-holding chore. Your man, especially on a rainy day, asks you to come help in the shop "for a minute." That means you will spend a slow age handing him nails, nuts, bolts, and assorted tools. Mainly, however, you hold things such as one end of an iron thing while husband operates on the other. Or steadying boards while he saws or handing up shingles while he nails. You can't sit down, you can't leave, and grunts are the only response to attempted conversation. Caution: Do not make the mistake of assuming an INTEREST in what your man is doing. Inevitably, this leads to comment and helpful advice from you, which is bound to be rejected—probably rather rudely. Try instead to invent new recipes in your head using weird combinations of ingredients. Serve it for supper as a shop-worn special.

*To achieve gate-opening flourish, pretend you are a star football tackler hitting the line.
**Country Women are masters of the Art of Assumed Nonchalance.

Sometimes you are called upon "just for a minute" to ride along while your man "brings in that cow with the foot rot." For some reason, this always develops into a triangle. One point of the triangle is your man—driving the pickup slowly in the wake of the limping cow. Another point is the cow, ambling down the road with an occasional glance from side to side. The third point of the triangle is you—on foot—jog-trotting along at the proper distance to keep the cow's mind pointed forward toward the corrals. Your job is to sprint like mad to cries (from the pickup) of, "head 'er off! Head 'er off!" when she tries to duck back. Then, at the precise moment, learned only through experience, you lope quickly ahead, swing open the gate, and gallop back to position. Your man, meanwhile, has been banging the sides of the pickup to promote continued forward movement of the critter. After cow and pickup have passed through the gate, close it. Then you may ride the rest of the way.

Sometimes you are privileged to help doctor a cow in the open without benefit of corrals or chute, especially on Sunday, the hired man's day off. Your man claims he's got to vaccinate "that number 35 heifer." You are requested to go along to "hold the vaccine gun."

"She's in the small pasture down by the creek and she's gentle," your man claims* (see footnote). "I'll just rope her and snub her to a tree." He further adds your only duty will be to hand him the vaccine gun at the proper moment.

Well, what else can you do? As a good Country Woman, you smile and make ready for a fun outing. Don't forget your hat and your leather gloves.

Admiringly, you watch as your stalwart husband quietly slips up on the chosen heifer and skillfully eases the loop over her horns. With an expert wrench, he ducks around a sapling, wrapping his end of the rope around it. At the other end of the lariat, that "gentle" heifer is going crazy—jerking, pulling, snorting, and bawling. Somehow, your man doesn't seem to hear or see anything.

"Hand me the vaccine gun," he requests, leaning hard on his end of the rope. You hand it.

"Here," he says, "hold this." You exchange vaccine gun for the rope end. "Keep it tight," says your man. (Don't bother to wonder why he thinks you wouldn't.) Recommended stance for holding lariat rope while critter occupies other end: Pass rope behind you as though you're going to sit on

*Any critter not actually a killer is termed "gentle."

it. This allows you to use all that ample posterior muscle as added leverage. So there you are, hanging onto a rope that has a captured, argumentative cow at its other end. Offer prayerful thanks for the wrap on the sapling. For a few seconds, the scene is a frozen pastoral tableau. The heifer faces you, head down, pulling mightily against the rope. If she'll hold that position, your man can slip up and stick her before she knows it. As your husband poises his needle, the heifer does a ballet leap straight up and comes down charging straight at you.

"Hold her! Hold her!" instructs your man. (You couldn't possibly NOT hold because you're too terror stricken to let go.) The cow hits the end of her tether at the sapling. You are saved. Then she sunfishes around and begins to circle the sapling UNWINDING that wrapped rope! Now is the auspicious time to move. Still hanging onto the lariat, zoom around the sapling rewinding the rope faster than the heifer unwinds it. Fortunately just as you are seriously considering abandoning the whole project, the critter plants her feet and hauls back on the rope. Immediately assume your semi-seated stance on the taut rope.

"Got her?" asks your man. You are safe in assuming this to be a rhetorical question. This time your husband succeeds in jabbing the heifer in her hip. She doesn't even move, only stares hatefully at you. Your man relieves you of your end of the lariat and frees the heifer, who trots five feet away and begins to graze, totally unconcerned.

"There, that wasn't bad," states your dearly beloved.

Assume your best air-of-nonchalance and agree.

A lariat, often, they say,
When practiced with day after day,
Can catch lots of critters
And give them the jitters,
But causes one's arm to decay.

INSTANT HIRED HAND

If you are a woman who lives in the country
You know you'll be called to assist
With swathing and baling and harrowing too,
The number of chores makes a list.

Though not very big and not at all tough,
You're *always* an Instant Hired Hand;
When something goes wrong, you're sure to find out
Just whether you've got any sand.

One morning you hear a request from your spouse,
(He's the man you married for better);
You know in your heart there's adventure in store,
And you get to help whatever the weather.

"Got a brockle-face cow in the South 40 field
With rot in both her hind feet;
I sure need a hand to help snub her down,
But she'll be easy to treat."

Your spouse takes his lariat off of its hook,
And gathers vaccine and stuff;
While you bring only a heart full of dread,
'Cause you know this job will be rough.

"Now, she's real gentle," so says your mate,
(Which means she's really a killer);
You bounce in the pickup down dale and up hill,
On a seat that has lost all its filler.

Down in the draw that is right near the creek,
The gimpy, old cow lingers there;
Then your hubby winds up, he drops a loop,
And war that moment's declared.

On one skinny sapling, he takes a tight wrap,
Then tosses the rope to you;
"Just keep her drawed tight," he says in your ear.
(Now really, what *else* can you do!)

So you park your rear on that raspy hemp,
Thus bringing your best to bear;
Your body's alert and it's poised to jump,
Real quick from here to there!

At the end of the rope, a cantankerous critter
Is pawing the ground on the fight;
You cling like a leech and mutter a prayer
That the lasso will hold her tight.

Your hubby is poised, his needle arm raised
To stick her in the rump;
"Now, hold her! Now hold her! Instructs your man.
As the old cow starts to hump.

She leaps straight up, sunfishes around,
Comes down a-circling back;
Around that tree she reverses the wrap,
At you she's making tracks.

With no time to think, you take off at a lope
And gallop the opposite way;
Now *she* unwinds while *you* re-wrap—
(It's a sight not seen every day).

You zoom to the east, she whips to the west,
You wonder which one will win;
You're about to quit when she plants her feet,
She's making an awful din.

As she glares at you, her eyes red-rimmed,
Your husband slips in close;
He sticks her hard with the trusty syringe,
Right in her hind-end roast.

When he slips off the noose and sets her free,
She commences to graze real nice;
"You see," says your mate, "I told you so;
We had that done in a trice."

You look at your palms all bleeding and raw
And think about killing the man;
Then change your mind when he adds real proud,
"And thanks, you're a darned good hand."

SECTION TEN

Branding

Wherein one learns how to deal with tally sheets, vaccine guns, and Mountain Oysters (or Prairie Nuggets, depending on where you live).

On branding day (the day when your ranch mark is burned into the hides of all those fresh-faced baby calves), you do double duty. Not only do you act as hired hand, but you're also chief cook, hostess, and runner-after of anything anybody has forgotten. Sometimes a day of indescribable hilarity, sometimes laced with small tragedies, branding day is always s a whole lot of work. As a thinking Country Wife, estimate how many neighbors and extra hands are coming to "help," add fifteen, then lay in mountains of provisions, because country custom says you must feed all those hands promptly at 12:00 noon. All food must be home-cooked. Fortunately, some of the neighbor Country Women feel it is their duty to help out in the kitchen.

While you nestle cases of beer and soda pop among the rocks at the edge of the creek, the men and women riders are furiously busy. Calves are cut away from their mammas, branding fires are started, ropers swing loops. Calves are roped and dragged, bellowing, towards waiting cowboys and cowgirls who grab the north and south ends of the poor little things and stretch in both directions Other cowboys leap to apply the hot iron, castrate the little bulls, notch the ears, and gouge out nubbins of horn.

You and the other women not smart enough to stay in the house (or if you're not a roper) are in charge of the vaccinating, the tally sheet, the disinfectant, the scour pills, the blood stopper powder, and keeping track of the de-horning spoons.

The watchwords for the Country Woman wielding any of the branding day implements are "Look out!" Exude confidence as you approach a calf captured firmly by north and south cowboys. Have the vaccine gun primed and ready. Pinch a bit of skin away from the calf's ribcage just under the foreleg. Thrust the needle in, making sure you don't poke through one side and out the other of the pinch, thereby squirting the vaccine in the air. Move quickly and try to keep out of the way of the man with the hot iron who is searing the brand into the calf's hide at the same time your are doing your needlework. The second you've finished injecting, pull needle out and turn nonchalantly away* (see footnote). While being nonchalant, do not dawdle.

*Caution: Although nonchalance is the key attitude, alert attention must be the Country Woman's prime focus.

Those two cowboys holding the calf will turn loose all at once. Being struck by flying hooves can take the enthusiasm out of the day.

About your person, hang a small notebook and pencil. Remember to tie these items on. Merely stuffing a notebook and pencil in your pocket won't do. At the wrong moment it will fall out and into a fresh juicy pile. Fish it out and use it anyway. As each critter is vaccinated, it's up to you to tally heifer or bull in your little book.

When you tally the bulls and heifers, you always get mixed up, but NEVER admit it. Always answer promptly and confidently even if you have to make it up.

Once or twice a lull occurs in the feverish activity. The men sag against the corral posts or flop on the ground, beer cans in hand. At this point, race for the house, make sure the big coffee urn is full and functioning, check the roasts, stuff a bushel of potatoes in oven to bake, mentally count the chairs around the dining-room table and decide some people will have to eat from laps in the living room.

Then race back to the corral where the thoughtful husband has saved you one last swallow of warm beer.

At the noon break, pick up the bucket of Mountain Oysters* (see footnote) and dash for the house ahead of the men. While they "wash up," load the food on the table. No matter how starved, the men will politely hang around outside in the yard till you ring the come-and-get-it bell. Once seated, absolutely no sound is heard for the first ten minutes besides the clinking of the eating utensils. When the first panic subsides, limited conversation begins to trickle around the table. By dessert, the morning's funny episodes and catastrophes have been repeatedly analyzed and all bachelors have been teased about their girlfriends.

Naturally, you and the other women keep the coffee pot pouring and serving dishes full. Between trips, you get to eat, standing up, in the kitchen.

*Mountain Oysters, sometimes referred to as Prairie Oysters: Those items separated from the bull calves. To be laboriously cleaned, egg-battered, crumbed, and deep-fried at a later date. (Freeze 'em till then.) Old timers claim they're delicious as well as having certain rejuvenating powers. It is not a good idea to dispute this claim.

To brand, you take an iron bar,
And heat it red hot in the far,
You throw down a calf,
And then with a laugh,
You singe his pore little rar.

In spring, in a ranching society
The ranchers cut calves with propriety
And then folks are fed
At a bountiful spread
Of oysters—the mountain variety.

THE LEGACY

My Grandmother gave me a gift today
(Now, I expected socks),
Instead, an apron, clean, but worn
Was folded in the box.

I know that women used to wear
Aprons to the floor,
Shapeless garments of muslin cloth,
Who wears 'em anymore?

My Grandmother doesn't know, I thought,
The modern way of things,
That wearing aprons and serving others
Is what enslavement means.

When Grandmother gently chided me,
Politely, I said I'd listen,
And as she spoke of the threadbare garment,
I began to see her vision.

"Rest your hand upon the apron,"
My Grandmother said to me;
"Listen to women who came before,
Women from history."

My fingers touched the roughened cloth,
I felt my spirit hasten,
As if the souls of those long dead
Were speaking through the apron.

I caught a glimpse of times before,
And I walked in ghostly shoes;
I fancied I was a frontier woman,
And wondered, would life I lose?

I held a baby as I marched
Beside a covered wagon;
And I was tired for the way was endless,
My weary steps dragged on.

Then howling warriors swept upon me,
Shooting from every side,
I ran till I thought my heart would burst,
There was no place to hide.

I slung the child apron-wrapped,
Not knowing where she fell;
The babe survived, and I am her kin,
Her name, like mine, was Nell.

Again I touched the worn, old cloth,
And became a prairie bride;
New-settled in a soddy hut,
At night, sometimes, I cried.

For fuel, I learned to follow bison
And picked up chips of dung;
I heaped them high in my muslin apron,
And fetched till my arms grew numb.

I carried thick, dark bricks of sod
Enfolded in my apron,
And learned to value prairie beauty,
As homeward I would hasten.

Once more I stroked the muslin threads,
And became a farmer's wife;
A skidding axe sliced through his leg,
Blood poured away his life.

To tourniquet the flow I tore
A strip from my apron hem,
Then harnessed the team and raced for town;
Death was foiled again.

Smoothing the cloth of Grandmother's apron,
I saw a farmhouse shelter;
A toddler wandered close to geese,
She chased them helter-skelter.

Those flapping, honking, pecking demons,
Gave the child a fright;
But right and left I thrashed my apron,
And drove those birds to flight.

Once more the vintage muslin apron
Made pictures in my mind;
I was my Grandmother in her youth,
My life was the Spartan kind.

When heifers broke the fence one day,
My apron closed the gap,
Till I could fix the fence with tools
And chase those critters back.

And I often sat by the back-porch door
And watched the children at play,
While snapping beans and shelling peas
In my aproned lap each day.

Then my Grandmother's apron showed me
Western women when,
At branding time or thrashing season,
They cooked for hungry men.

I saw them in their kitchens,
Roasting, baking, heating;
Then folding hands across their aprons,
They proudly watched men eating.

"Have some more, there's plenty here,
Now, fellas, don't be shy!
More coffee, Joe? Have some cake!
Here's a piece of pie."

"This apron is my gift to you,"
My Grandmother said to me,
"Woven with truth and women's power,
It is your legacy."

I reached across my Grandmother's apron,
And touched her gnarled hand,
And felt the souls of all the women
Whose courage forged this land.

I thanked her softly for my present,
In tears, and filled with pride,
I greeted the shadows of aproned women
Walking by my side.

PART TWO

Good Old Summertime

Summer: That time of year when one is impossibly busy. The ranch or farm bursts with life, and company from the City descends in voracious hordes.

SECTION ELEVEN

Summer Irrigation

Wherein one learns what it is to cope with hired irrigators, husbands who work dawn till dark, and what potions and lotions to apply to cracked feet and blistered hands. One also learns about irrigating boots and how to reply to those dudes who want to know if you're goin' fishin'. And wherein one also learns how to hold the irrigating shovel.

"As early as possible" is your husband's motto regarding irrigation. That is—early in the spring and early in the morning. Should a farm or ranch happen to utilize the technological upscale pipe or sprinkler irrigation systems, those folks can count their blessings. At your modest spread, everything's still done the old-fashioned way—flood irrigating.

Some ranch operations retain a year-round hired man who is then available to help irrigate—if he doesn't quit. Since yours is a family operation, your man enlists your Instant Hired Hand services. As summer develops and daylight extends, more and more time has to be allotted to irrigating. Finally, your man has to go to town to hire a hand. He may come home with anything from a broke-down sheepherder to a high school kid. In any case you exchange some (but not all) your irrigating privileges to devote more effort to the care and feeding of hungry men. This is not necessarily an improvement in status.

Irrigating and haying are two ranch activities that increase a man's appetite threefold while slimming down that bit of a paunch he developed over the winter. Sadly, while your husband slims down on the extra work, sweat, and food, you are apt to increase in girth.

Whether you have old, young, lazy, competent, or incompetent irrigators, all of them sweat. As a good Country Wife, encourage bathing and salt intake. With every meal, set out a giant bottle of salt tablets. The men will swallow these readily. Take some yourself also.

Those men who do not believe in bathing become pungently noticeable. Be very careful about hints to bathe. If the man feels insulted, he may quit. And your husband prefers a stinking, good irrigator to doing without help. A plentiful and frequent application of room spray may be all you can do.

As a thoughtful Country Wife, fill several canteens or water bags so the men can carry drinking water. Likely they will forget them. When thirsty, an irrigator drops to the ground and sticks his face into the ditch, just like you see in western movies. For some reason, irrigators never fall ill from drinking ditch water.

When your man strides forth in his seven-league irrigating boots, his shovel balanced on one shoulder, and his hat pulled low over horizon-seeking eyes, he looks John Wayne impressive.

The picture you present is a trifle less romantic. Irrigating boots are rubber monsters that chew your poor burning blistered feet relentlessly. The tall rubber things extend halfway up the thigh, but are worn flapped over to reach just below the knee. You're forced to walk slightly aspraddle. Wearing your tacky, oldest straw hat, packing a shovel, and walking spraddle-legged, you look a little like a chubby peasant off to the rice fields.

Probably one of the most important things to remember about irrigating is how to hold the shovel. When walking from place to place, the proper carrying method is to angle the shovel across one shoulder. At intervals, you must dig little gullies to divert the water to spread over the pasture. Or you have to spade up some sod to block the gullies because that part of the field is wet enough. After gully digging, you are required to spend a moment thoughtfully studying the progress of the flow. At these moments, use the "Water Watcher's Stance." Hold the shovel handle at the top end by both overlapped cupped hands about chest high. Angle the spade end out in front of you to rest in the dirt. Spread your feet slightly in the "at ease" position. Viewed from the side, it appears the shovel handle is propping up your body. The "Water Watcher's Stance" is also the ready-for-action position. You are ready to jab, spade, or scoop wherever water needs to be diverted or dammed.

As you trudge along following the ditch where it meanders across the hayfield that borders the river, watch out for two predators—rattlesnakes and weekend fishermen. Both are harmless if dealt with decisively. Every year one or two rattlers sun themselves right at the edge of the culvert where it carries the ditch water under the road. Do not hesitate or debate on the ecological balance of nature. Kill it. To kill a rattler, bang it viciously with your shovel. When it uncoils and tries to slither groggily off, jab down hard with the sharpened shovel edge, and cut off its head* (see footnote). Cut off the rattles as soon as you're sure the thing is dead. Your oldest son is collecting them. Bury the head and the body, too, unless your son wants to tan the hide to make a hatband.

Sometimes as you cross the county road, a shiny car, obviously from the City, slows to a stop. A shiny man leans out and inquires how the fishing is. Remember it is not legal to do away with tourists or fishermen just because they're in season. Smile your simplest country smile. Then direct him to a fishing access on the river, instruct him to close all gates, and to refrain from parking or picnicking in the hayfields. Then pull out your fisherman's log and take his name, address, and license number. If he gets testy about giving his name, he'll roar off in a cloud of dust. Which means you wouldn't have wanted him anyway. Those who are friendly enough to chat and part with their names often turn out to be fine folks, and will even give you part of their catch.

On the long march home, you wish you hadn't forgotten your water bag. At the house, slip gratefully out of those nasty rubber boots. All summer long, both you and your husband will suffer from blistered, cracked, burning feet unless you take precautions. Daily, apply Bag-Balm liberally** (see footnote). On top of that, apply cold cream. In addition, powder the insides of the boots. Lastly pour a glass of iced tea from the big pitcher stashed in the refrigerator. Sit down and read for ten minutes before you get up to start dinner.

A well-oiled irrigator named Durbin
Found falling in ditches disturbin'
He lay there and wallowed,
And gasped as he swallowed,
"This ditch water sure needs some bourbon!"

*Several blows may be required. The darn things are tough.

**Bag Balm: A soothing salve used to keep the milk cow's udder and teats soft and smooth. Great as a hand lotion as well. May be purchased at the local Farmers' Co-op.

SECTION TWELVE

Haying

Wherein one learns to accept as normal the preparation and serving of mountains of food to hay crews and how to haul a snack to the fields, plus complete advice on operating the tractor, the swather, the baler, and the mower and make it appear that you know what you're doing. One also learns that some, new-fangled farm machines have enclosed, air-conditioned cabs, TV, radio, and more floor space than your entire house.

Haying is that picturesque country activity indulged in by ranchers and farmers every summer. A clear, blue sky arches over ripe, yellow grain fields and lush, green-gold hay meadows. City dwellers drive up and down the country roads in their air-conditioned vehicles and point at the sweaty men and growling machines.

Foot by foot, meadows are mowed and the shorn grasses raked into windrows. Or if the swather is employed, the grass is cut and swept into even windrows without being raked. Then along comes the baler to eat the windrows and spit them out as hay bales. Eventually, row upon row of bales dot the crew-cut fields. After that, the farmhand* (see footnote) bucks the bales onto those beautiful haystacks the city folks admire as part of the summer scene.

Sometimes as a preliminary to cutting or swathing the meadows, your man will hitch the mower to the tractor and ask if you have time to "trim the ditches." Of course you have time.

Mowing the ditch edges first helps more important and more expensive machines gauge how near they are to the ditch edges. A mower is a long toothy thing that sticks out sideways from the tractor. Sharpened teeth lie in a double row horizontally to the ground. As the tractor moves, pull a certain lever and the teeth gnash back and forth sawing through the tall grasses. There are several things you are not supposed to do when operating a mower. You're not supposed to slice rocks, small trees, or fallen logs, or jam the end into the side of the ditch.

At other times you may be asked to operate something called a "side delivery rake." You are inclined to interpret that as meaning rakes give birth to little rakes on one side. That is not true. And don't mention the thought to your husband, especially if he's behind schedule.

*A farmhand is not a person.

The rake is employed on those areas where a mower has cut the entire field, generally a small patch where the big swather couldn't maneuver well. The rake scoops up the mown hay and "side delivers" it into windrows. It is considered OK for the Country Woman to operate a rake.

Rarely is the Country Woman called upon to run a swather* (see footnote). Swathers are fun because you sit high, grandly chomping down a whole field. The thing has great, huge, circular teeth that look like the bristles of a super carpet sweeper. A swather cuts and mashes the grass and sweeps it into windrows all in one operation. The mashing process "conditions" the hay so that it cures faster. However, the beast is apt to suck up and mash dead skunks, pigeons, and other deceased varmints. When the undiscerning baler comes thonking along, it picks up and squeezes grass and varmints all together. (Breaking open a bale containing a rotted animal can be a pungent experience.) With the swather, avoid all the pitfalls you should have avoided with the mower and the rake. In addition, it's very important not to drive into the bog. A swather stuck in the bog is guaranteed to drive a good man berserk. Whether you could help it or not has no bearing.

In fact, a bog-stuck Country Woman better have a good excuse. The thinking Country Woman has a stock of creative explanations prepared such as: "Had to avoid a boulder, which would have broken a wheel;" or "Had to avoid a kill-deer's nest of young ones;" or, best of all, declare you were "following the exact line" set out by your husband to avoid falling in the bog.

When mowing, swathing, and raking are completed, two or three days must elapse before baling to allow the hay to "cure" in the sunlight. If it rains, cursing and teeth-gnashing ring out across the land.

The machine thing with the rhythmic bang-thump-bang is called a baler. Growling like an angry dinosaur, it picks up the windrows of grass and sucks them into its craw, then spits a string-tied bale of hay out its back end. If you are drafted to run the baler, your neck will develop an awful crick from trying to watch fore and aft all at once. Watch ahead for rocks, pitfalls, and deceased or live varmints. Watch behind so that you know the baling twine is feeding through and tying the bales properly. The tension should not be too loose or so tight the string breaks. If a choice is necessary, it is better to bale a varmint than to break the darned string in the middle of the operation.

*You only operate the swather in cases of severe emergency such as your man having two broken legs. Men adore swathers and guard them jealously.

You are always willing and eager to operate the machinery anytime. From your perch on top of the world, you have peace to compose letters to friends, to sing songs, to develop a new pattern, mentally write advice to your congressman, and watch the bald eagles swoop over the river. Be careful to pay some attention to where you are going. It's during your most beautiful thought that the bog will suck you under.

For a Country Woman, the absolutely most important thing to learn about machinery is the correct attitude. Approach and demeanor carry the day. Dress with a certain jauntiness. Jauntiness intimidates machinery. Wear your new straw hat and leather driving gloves. In order to avoid a patchwork-quilt tan, do not wear short-sleeved, open-throated shirts. Haphazard tanning ruins the look of your best white sleeveless low-cut party dress. It makes your arms—brown to the vaccination scar—appear as though mismatched with your shoulders, while your neck and throat look as if you're suffering from a serious liver problem. High-necked sleeveless blouses are recommended, so that at least your face and arms tan evenly. On a hot summer day, the best garb of all is a bikini, if you can get away with it. Perched atop a machine, clad in your orange flowered bikini, work boots, straw hat, and leather gloves, you present an arresting picture. Arrested is one of the things that will happen to you, your husband claims, if you don't wear something additional!

There is one chore everyone feels is more compatible with your talents. It's called turning-the-bales. Bale turning is particularly suited to you, say the men. Don't argue. Just advance to the field where about a million heavy bales dot the landscape. To make it easier for the stacker to scoop up the bales, they must be turned at right angles to the farmhand teeth. That part is pretty simple. It becomes more difficult when you have to turn the bales all the way over (a fresh bale weighs from sixty to one hundred pounds) so that the down side is now up so that the hay won't rot from the morning dews soaking into them.

While turning bales, wear something cool, but not your bikini or you'll be spiked to death from the spears of hay. Wear long sleeves, jeans, leather gloves, a straw hat, and tie a jug of water to your waist. Choose your own style of bale turning. Some folks like to start on the perimeter of a field and go in ever-diminishing circles. Others prefer to march back and forth, gradually evening up the field. Upon reaching the center of the field, or your last bale, whichever comes first, sit on it, and have a big draft of water. Contemplate the bales that you should have been counting as you went along. If you've lost track, just decide on a number you think is close. WARNING: Bale turning can be hazardous to health. If you live in snake country, it is possible some of the creatures will seek out the cool underside of a hay bale for a siesta. In addition to the water jug you carry, tie a stout, forked stick to your belt. Fix it so you can grab it quickly. Usually the snake, startled at the interruption of his snooze, will try to slither away from you. Immobilize him with your stick, then dig out your pocket knife and cut off his head and his rattles. Ignore any squeamish feelings about executing the snake. Better a little squeamishness than to wonder where it wandered off to—perhaps stalking you.

The cruelest part of a haying operation is the stacking...for you anyway. Some very robust Country Women can maintain the lift, throw, push, and pull pace of stacking with no ill effects. You, on the other hand, barely refrain from screaming as the pain increases in your back, arms, legs, and feet. Even your eyes and teeth hurt. But you climb once again to the top of the stack because you'd die before admitting you are past your endurance.

Bale-stacking technique is not difficult to master. Your man bucks up the bales with the farmhand, a wide thing applied to the front of the tractor. Several five-foot teeth set about twelve inches apart extend from it like bristles. The monster looks like a huge comb big enough to groom a pine tree. It scoops up a bunch of bales and carries them over to the stack atop which you stand. At the push of a lever, the teeth rise straight up to your level, then tip forward to dump its load. Grab the scattered bales (don't fail to bring along your bale hooks) and shove and push and lift into place. Stack one direction one row. Change to the other direction on the next row. Alternate directions all the way. No matter how tired you get, be alert when the farmhand growls up with a load. Being buried under an avalanche of heavy bales is truly awful. Besides, your man considers it time wasted to stop and administer first aid.

Both the haying crew and the stacking crew consist of your husband, a possible hired man, your oldest son, and, of course, you. Throughout all haying operations, there runs a terrible sense of urgency. The hay has to be cut before it's too ripe, then it's got to be baled before it rains, then it's got to be stacked as soon as possible.

As a good Country Woman, your main job is to keep the food (food production is still considered your primary job), drinks, and salt tablets going.

Often ranchers will hire a hay crew or one rancher will help another. No matter how it's done, you will have extra ravenous mouths to feed. Basic rules are simple. Prepare plain food and lots of it. Start each day by sticking a roast, a giant stew, or several chickens (not live ones) in the oven. At the right time, add potatoes and always make buckets of gravy. Salads and vegetables of any kind will do as long as they're not fancy. The men won't eat those dinky things with the sauces, but they'll eat anything in Jell-O. Provide plenty of bread, butter and gallons of coffee . . . always. Cool drinks are acceptable as an addition, but never as a substitute for coffee. You can't miss with ice cream and pie or cake for dessert. Buy the ice cream in five-quart plastic buckets at the grocery. Always save the plastic buckets to fill with your own homemade ice cream when time permits. The buckets also make handy containers for holding nails, slop, or collecting eggs.

Mid-morning and mid-afternoon are key moments during haying season. As a noble, good, loving Country Woman you ride to the rescue of the thirst and hunger-crazed men at those times. Which means you cheerfully interrupt what you're doing to cater snacks to the fields. As you arrive toting your load, the men observe you immediately, but always make another round before pulling to a stop. This is to indicate they're ALMOST too busy to halt for a snack. Amid a great deal of brow-wiping and complaints about the heat, three-fourths of a cake, a jug of ice-water, Kool-Aid, and a big thermos of coffee disappear. Then, at a secret signal, the men all rise and thank you profusely; each pulls out a used toothpick from a shirt pocket or hatband and remounts the swather or the baler or the rake or up on the haystack. This is all done in one concerted movement, leaving you standing or sitting among a scrap pile of coffee cups, cake crumbs, and thermos jugs. You will have a distinct feeling nobody even knows your name. However, bravely tidy up and trudge home because it's almost time to start dinner or supper* (see footnote).

Bales come in other shapes than the small oblong units you've spent all that muscle power stacking. You are filled with justifiable pride, but on the neighbor's ranch they're still haying using machinery that gathers the hay into gigantic round tubes, flat on both ends. The big round bales weigh a ton each. Not even your husband can lift and throw the things. No, a gizmo with robot-type arms mounted on a flatbed takes care of hoisting the bales. Once on the truck, you can drive to the field, push some levers, and off will roll the bale.

*On the days when you are drafted to run a machine, prepare snacks ahead of time and take them to the fields with you. A snackless day is a day of moaning in the fields.

Then you jump down, cut the strings with your handy pocket knife, and the bale is supposed to unroll. Except it doesn't always. Surrounded by hungry cattle who have no qualms about running over the top of you is not a comforting experience.

Whichever method or style of bale, it's the cuisine of choice for cattle, sheep, hogs, and any other more exotic farm or ranch animal, not to mention free-loading deer and antelope.

The hayhand fell off of the stack
And fell with a terrible whack
But his wife braved the fescue
To come to his rescue
And toted him home in a sack.

SECTION THIRTEEN
How to Keep Cool While Haying

Wherein one learns the fine art of driving a tractor, baler, or swather while wearing as little as possible under a blistering sun, and wherein one also learns passersby are highly entertained even if you're neither young nor nubile.

What with drought, forest fires, and relentless, merciless heat, haying isn't easy—which is like saying that having all your teeth pulled without Novocain is a pleasure, or that you're pleased when the guy on the ditch above you steals your irrigating water, or that your sheepdog has no bad habits.

Seated in the tractor cab of swather or baler with the sun beating in can be worse than torture because nobody's making you do it. You could have chosen a different life—maybe swimming with sharks, or skinning lizards, or volunteering to write speeches for a politician. (Those agriculture persons who own tractors with air-conditioned cabs, radios, and tea service every two hours are not to read this. I'm told the very latest tractors have tiny TVs and cell phones, ice-machines, and beer—well, maybe the beer is an exaggeration.)

The norm for regular, patched-jeans, large-debt, aching-back ranchers and farmers is used equipment lacking all luxuries. A padded seat on a tractor happens only if the operator cuts a piece of foam and duct tapes it in place. If there's a windshield, it's usually just the frame, because the glass has long since fallen out. Operating the vehicle is a little like you might feel if somebody put you in a jar and shook it; however, the motor noise is quiet—if you wear earplugs.

Jill and Marcia, two country girls desiring to earn money for college, hired on to put up the hay for a neighbor who hasn't had a new piece of farm machinery since forever. Marcia chose the swather, which sported a cab that had a roof—barely. The side windows were stuck shut, and air conditioning was what happened when a raven flew past flapping its wings.

Jill's baler tractor was a 1948 Ford with attitude. In the searing, brain-shriveling heat, keeping cool was going to be a country-girl challenge.

"We'd better carry lots of water," said Marcia.

"Yeah, and wear as little as possible. It's gonna be a killer out there."

Jill donned personally cut-off jeans and a shirt that had no sleeves, no back, and very little front. A broad-brimmed straw hat protected her head, sunglasses shielded her eyes, and a ton of sunscreen covered all exposed skin, but she wore regulation cowboy boots in case of snakes. That particular hayfield lays adjacent to the state highway. Whenever a round put her parallel to the road, many passing motorists, especially males, waved. Jill, being a friendly girl, waved back.

Marcia, driving the tractor with the stuck-shut cab windows, began having hallucinations as the temperature grew hot enough to turn rock into lava. A camel might have been happy, but Marcia definitely wasn't. She drank water, then she drank more water. She perspired. The sun beat down. She could feel herself frying. If she didn't do something, she was going to have a stroke. She was sure of it. Then she had an idea.

Pulling to the far side of the field she peeled down to her lacey undergarments (even country girls like lace in certain items). Her water bottle had one of those spray-nozzle tips and Marcia proceeded to wet herself down. A thorough dousing lasted approximately two and a half rounds before she had to re-spray.

Eventually Jill and Marcia crossed trails. They paused.

"You keeping cool?" shouted Jill above the motor noise.

"Yep," yelled Marcia, "I'm spraying."

"Is that so?" Jill hollered back. "Well, a little praying can't hurt, might help."

"No, no, I said spraying, not praying," shouted Marcia waving her spray bottle up and down and back and forth.

"And blessings to you, too," screeched Jill as she drove on.

SECTION FOURTEEN

Moving the Cows to Summer Pasture

Wherein one learns that you, as an incompetent Instant Hired Hand, can lose a calf, but it takes an Experienced Cowman to lose a bull. One also receives lessons in correct vocabulary to employ while on a cattle drive as well as the reason for starting the drive before dawn.

Sometime after the grass has greened up, your man squints at the sky, scuffs the ground, and eyeballs the critters. In that mysterious, intuitive way, he ascertains it is time to move the cows. Which means all the mamma cows, along with their children, must be gathered and pushed over the hills to the upper section.

On the evening before the magic day, your man casually remarks, "You have time to help ride tomorrow?" Of course you have time. You wouldn't miss it no matter what. Even pregnant, you can ride till you're six months along provided you don't fall off. Long before dawn, about 3:30 a.m., rise and whip up a huge breakfast. In there somewhere make about a thousand sandwiches, several jugs of coffee, and toss in a bunch of apples, cookies, and/or cake. After breakfast pack the lunch. You'll have plenty of time because it'll still be dark for quite awhile. Country men rise in the middle of the night in preparation for a cattle drive. This is a ritual developed by menfolk to the tune of noble grunts and groans, not to mention creaking bones. The men stagger forth to sit around the breakfast table drinking coffee and waiting for daylight* (see footnote). The food, the coffee, and some horses are loaded into the truck and unloaded at the Otter Creek corrals, three miles across the hills. From here, your husband, your son, the hired man, sometimes a neighbor or two, and you will begin the drive.

You haven't been on a horse since last fall's roundup and both you and your mare could use some preconditioning, especially around the hips. However, as the sun rises high enough to be able to discern cow shapes from tree shapes, the two of you make a charming western picture against the morning skyline.

You are directed to gather "those cattle over there" and "don't miss any of the coulees." "Over there" encompasses a half section of sagebrush hills and rocky gullies. Pull your special cowboy Stetson down snugly, but at a flattering angle over one eye, and commence. Ride up the ridges and down

*Waiting for daylight is a habit afflicting only men. It has something to do with being tuned in to the universal male rhythm of life—if you want to get silly about it.

the draws, swing into the brushy bottoms, and check behind every outcropping of rock. Gradually you will gather several pairs of cows and calves. Patiently head them up the draw where the men are gathering critters from other directions.

Look to the right and admire your man as he expertly turns a spooky calf back to the herd. Just then that wild-eyed little brockle-face calf in front of you ducks aside, zips around your horse, and gallops madly down the coulee—away from the herd. Kick your mare into a lope and try to circle wide around the calf. But the critter stays just ahead of you. As he suddenly disappears around a small hill, boot the mare into a gallop and cut over the top. If your strategy works, you will be ahead of the calf. He stops. You stop. Gradually and slowly ease toward the calf. If you can get him pointed the other way, he will tear back to his mamma. It's a matter of pride, honor, and fear to not lose that calf* (see footnote).

As the calf stares at you, his eyes roll up as though having a seizure. Jumping a foot straight up, he lands running—in the wrong direction. Do not hesitate. Spur your horse. You've got to cut him off before he reaches the fence bordering the road. There's an underpass there used by the cattle to pass from one side of the pasture to the other. If he ducks through that he'll have a section and a half of unimpeded territory. Your poor chubby mare begins to wheeze and so do you. But you're both game. The mare, sensing victory, puts on a burst of speed. The calf, sensing freedom, also puts on a burst of speed and whizzes through the underpass. Ears flattened in anger, your wheezing steed pursues. The fact that the underpass is barely higher than the back of a horse could be a worrisome factor if you thought about it. This is not the time for thinking. Cleave closely to that bobbing horsey neck and hang on. On the other side, open your eyes and regain control of your mount. At this point the calf is half a mile away, so let the damned thing go. While returning to the herd, use the time to prepare a good strong story; practice your nonchalant air and don't ride too close to your husband for awhile.

Don't feel, no matter what your man implies, that you're the only one to let an animal slip away. As the trail gets longer and hotter, that big, black bull with the froze-off ear heads for the brush. Your man pushes him out of the creek bottom three times. Where a fence line sweeps over the little butte, the old bull disappears. While the rest of the riders ride on, your man rides to

*When a man loses a calf, it's a matter for comparative discussion about how snaky the little cuss was. If YOU lose a calf, it's due to a lack of attentiveness, ability, and brains.

find the bull. Half an hour or more later, your husband, the Experienced Cowman, rejoins the group of cowboys and cattle—with the bull. Do not ask any questions. Eventually you will learn the true adventure tale of how the bull escaped into the thorn bushes. The tale will grow with the telling until the bull develops into an incredibly foxy old so-and-so. How to capture such an outstandingly clever bull becomes a hot subject for quite some time.

Whenever a cow critter decides to go her own way, or becomes balky, or hangs back, she is urged forward on the path of righteousness with some well-chosen words. These words describe her ancestry, her moral fiber, and her attitude toward life. Most of the words need only four letters to spell them. As a good Country Woman, ignore all of these verbalizations. The men are barely aware of the pearls falling from their lips. Rich descriptive vocabularies are always part of a cattle drive. You don't condone the language, but are not above an occasional sizzling comment yourself. The time the old, horned cow bunted your horse and caught you in the kneecap, your man, the hired hand, and two neighbors regarded you with respectful awe.

At the end of a couple of hours of riding, cows and tired calves crest the last ridge and spread out over the new seeding your man planted last year. The green grass almost brushes the cows' bellies. It's a fine sight to see next year's income happily munching away.

Dismount along with the men and take a few moments to loll and watch the calves mother up with their correct mamma. Although the men hunker, it's better for you to put your knees on the ground and sit back on your heels. Hunkering is a manly art not recommended for women* (see footnote).

Lolling time over, the men make noises indicating they're about to mount up. This is the signal to crawl in the saddle and get a good, fast head start. Lope the mare over the hill and descend to the creek bottom. It's the only brushy spot for miles and your physical problem is becoming more and more urgent. Toilet technique while on a cattle drive in open country with male companions can be painful. Usually, when you begin eyeballing every little rock and sagebrush, one of the men chooses that moment for friendly discussion. (They've ignored you for over three hours up to that point.) A selected team of Country Women researched this particular subject. The objective conclusion reached was that male persons simply don't HAVE to go as often as female persons. In the event that a male person

*If you try to hunker, your legs cramp and your feet go to sleep. Rising smoothly becomes impossible. You're reduced to rolling forward on hands and knees, then clambering awkwardly and painfully upright.

DOES have to, he merely rides apart, stops his horse and dismounts to adjust the cinch. Turning casually away (on the far side of the horse), he spends a moment scanning the horizon, his clear eyes ostensibly searching for critters while he ponders the meaning of life. In just a little while he apparently finishes scanning and pondering, mounts up and proceeds with the cattle drive.

Things aren't quite so easy for female persons. One coping method is to take along a camera. When the moment becomes desperate and you can't get rid of a male companion, announce you want to take a picture from "over there." Tell him to "ride on" so you can obtain a cowboy rider in the photograph. Then dash for that clump of sagebrush. If you can stand it, take the picture first* (see footnote).

When you finally ride down that last draw and see the Otter Creek corrals spread out below, you decide you'll live after all. Everybody dismounts, bone tired, dirty, sweaty, and played out. The men tie up the horses in the shade and flop down alongside. You, of course, go to the truck and begin to unpack the sandwiches.

After a huge intake of sandwiches, coffee, and dessert, the whole crew revives slightly. Relax blissfully flat on the ground and listen as the men describe and explain and remark about each cow, the remarkable growth of the new grass seeding, and how the locoweed isn't so bad this year. Contentedly pull your hat over your eyes and slip peacefully off to sleep.

*It doesn't matter whether there's actually film in the camera.

A GOOD HORSE, A GOOD DOG, AND A TALL BUSH

A cowgirl has a heap of fun
A ridin' for the brand;
She rounds up cattle, ropes them calves
As good as any hand.

She wrassles doggies on the ground,
And also vaccinates;
She cuts the bulls, tattoos the ears,
And keeps the tally straight.

And when the cowboys take a break
To have a pop or beer,
A cowgirl also drinks her share,
Then faces wrenching fear.

This cowgirl's had six cups of coffee
And juice to cut the dust,
With added beer, she's into pain,
To go is now a must.

For it has been five hours since dawn,
She's way out on the prairie;
With not a twig or tree in sight,
The pressure's getting scary.

Now a cowboy does it simple style—
Just stands there by his horse,
A pondering life's meaning while
Nature takes its course.

But a cowgirl has three things she needs
While working way out West,
For riding chores, a darned good horse
Will help her do her best.

And second on the list, a dog
Who'll jump to her command,
And heel the strays and line 'em out,
A pup with lots of sand.

But happiness is when she finds
A tall and leafy bush
To shelter her and block from view
Her unclothed cowgirl tush.

SECTION FIFTEEN
Veterinary Care

Wherein one learns how to haul a sick critter to the vet. One also learns what to do while the veterinarian delivers a calf by Caesarean section.

Survival of farm animals in pre-vet days must have been sketchy. It's the worst kind of discouragement to watch healthy, bouncy calves, piglets, or lambs sicken and die. Often the little creatures expire seemingly for no reason at all. Sometimes the disease that carries them off can be identified and sometimes not. A common killer of baby bovines is calf "scours"* (see footnote). Before penicillin and antibiotics, farmers and ranchers tried a remarkable variety of cures. Only the strongly constituted survived. When a critter got sick, it got better, or it died** (see footnote).

A veterinarian serving a rural community is endowed with incredible courage and hardiness. At any hour and in any weather, he will come to a critter's aid. Unlike people doctors, vets still make barn calls. However, they do charge mileage.

When an animal is sick, it is more convenient (and cheaper) to take it to the veterinarian's office where he has proper tools and equipment. Hauling a sick critter to the vet frequently falls to you, as chief town-goer. This is especially true if it's a baby animal.

On a snowy, cold morning towards the end of winter, it becomes necessary to haul a sick bull calf to the doctor. In the back of the station wagon (the vehicle used to haul kids to the bus stop at the county road), your husband spreads a large piece of canvas. Since you'll be gone for a long spell, leave a pre-cooked meal available if possible. Otherwise, for a meal, the men will have bread, butter, and huge bowls of dry cereal, as the word "stove" to them means "crippled up." The bowls will be piled in the sink, pathetic testimony to their avowed inability to "cook."

You, with domestic chores under control, climb into the car and start toward town. It's best to keep in mind your husband's last words, "Drive carefully." It has begun to snow, the flakes coming down in soft, wet, clinging globs. The little bull calf lies behind you flat on his side, twitching and shivering, eyes showing white. All goes well for several miles. The calf has

*Scours: A general term for that loose bowel condition that comes in a variety of colors. However, mustard yellow is usual.

**An old-time remedy for scours from The People's Home Library, circa 1916: "Take a piece of bright colored yarn and tie it around the calf's tail close to the root."

been too sick to do anything but lie there, but suddenly he begins to blat and flop his head about. Speak soothingly. The soft tone won't do much to calm the calf, but it will make you feel better. As you round that rather sharpish curve halfway to town, the calf rolls sideways. The motion stimulates him to activity. He struggles to his feet. Wobbling and lurching wildly from side to side, eyes rolling, tongue lolling, the calf emits a horrendous bawl—right in your ear. Do not let this bother you. Concentrate on driving, alternately viewing the calf's actions in the rearview mirror and peering through the snowy windshield. As the calf lunges over your shoulder, bawls in your ear, and drools down your back, reach with your free hand to push him back from time to time. Don't, however, startle him. At all costs, keep his front end facing you. Creep the remaining distance to town, one hand lapped over the back of the seat, ready to push on the calf's brisket when needed. The last few miles become an exercise in bearing pain. Circulation has ceased in the cold fingers of the hand gripping the steering wheel. Neck and shoulder muscles are steel bands painfully tensed against a possible forward rush of the calf. The arm and hand half twisted over the back of the seat is about to dislocate at the shoulder* (see footnote).

Eventually, you reach the sanctuary of the vet's office. Gratefully pry yourself loose from the steering wheel and try to reorganize kinked muscles. The doctor, a kindly intentioned man, asks if you have a problem. Refrain from comment on the obvious as you hold the door while the doc attempts to capture the calf. The reluctant and terrified patient is transported by hand to the treatment room. Once the doc flops the calf onto the table, you are called upon to assist by holding things down. That means four legs, a head, and a flying, fecal-saturated tail must be kept under control. Drape yourself over the calf, octopus fashion, taking care the tail doesn't pull free to flap in either yours or the good doctor's face. While you maintain a four-corners control, Doc administers medication. The sick calf is usually dehydrated and must have something intravenously. While the cheerful veterinarian arranges the intravenous set-up with its tubes and bottle, maintain your close relationship with the calf. Try to achieve a mutual trusting rapport because Doc expects you to hold the calf still while he inserts a giant needle in one of its veins. Should you suddenly begin tossing and jerking, there's an even chance you too may receive something intravenously.

*During this sort of medical emergency, all Country Women develop a third arm and hand. No one knows how.

Whatever treatment is applied, it smells like rotten effluvium. The nasty odor clings to clothing for days. Finally, Doc is finished. He scoops up the calf and carries him to the car while you trail behind carrying a bag of those same smelly potions to be administered at home. The short parade is met at the door by Mrs. Super Country Woman, her face arranged in greeting. Actually, she greets Doc, who (still holding the calf) courteously responds to her questions. She tosses you a conversational morsel by mentioning how well you look and that HER husband wouldn't expect HER to haul a sick critter to town.

Smile and pass on by. She'll never notice, for awhile, that glob of fecal material clinging to the back of her pale beige coat.

Sometimes the veterinarian must come to you, particularly if a heifer has to have a Caesarean and it's too late to haul her to town. Your husband prepares a clean stall in the barn and spreads around a generous layer of fresh straw.

When Doc arrives, the heifer is suffering, her breathing labored, her sides heaving. She must be encouraged to lie down. Using a system of rope loops, the two men throw her down, left side up, as gently as possible. In fact, they don't even cuss much around a cow in labor.

The cow's front feet are tied together and then the rope anchored to the side of the pen. The action is repeated for the back feet with the holding rope then hitched to the opposite side of the pen. When the cow is trussed so she can't kick, there remains another portion of her to control, especially if she's fidgety. Your husband and the veterinarian look at you and state, "OK." Which means you now move to the cow's head and sit on it. (It's helpful to remember that cows only have teeth in their bottom jaws.) Sit facing the operation zone and more up on the neck than actually on the head itself. You're fortunate if she's a cow without horns. She can't do much in this situation, but every time you look around, she's staring at you. When she catches your eye, she moans and her eyeballs roll wildly.

As soon as the Novocain takes effect, the vet makes a long incision in the mama cow's side. While the doctor works, you don't have a lot to do except become educated* (see footnote). Your man, meanwhile, is preparing the resuscitator, which is a thing that blows oxygen into a calf's lungs while sucking out phlegm and mucous from its throat. As soon as the newborn emerges, the doctor and your husband work quickly to get the little

*Attending a Caesarean is not a good idea if you are squeamish or pregnant yourself.

critter breathing. If no resuscitator is available, alternate methods include draping the calf by his hind legs over the top rail of the stall, or swinging him, nose down, while holding him up by the back legs. While the holding and draping operation is going on (fortunately you don't have to do that as you're too short—unless of course, you're the only other person there), your job is to insert a tickling piece of straw up the calf's nostril, which causes him to sneeze and thereby blow out the phlegm. Naturally, you have removed yourself from the cow's neck to engage in the tickling procedure. Once that calf is out of her, the mother seems to relax a bit.

When the calf is breathing well, you and your man rub him briskly with a clean rag and work his front legs back and forth. After he's dried off, help him lie down so his feet are under him. In order to avoid danger of those fresh, new lungs filling with fluid, don't let him lie on his side.

Finally the operation is completed. The calf looks fine. The mother is relaxed, the gash in her side stitched tidily together. Your husband and the vet release the holding ropes and untie the cow's feet. She pops up as though nothing had happened, walks over to the newborn child, and beings licking* (see footnote).

All the people leave the pen. Scrape the crud off your coveralls, thank the vet, and invite him to the house for coffee. He thanks you and states he'll be along in a minute. As a good Country Woman, take your cue and leave. Your husband stays behind to help Doc gather his gear. In the course of gathering, your spouse extracts a bottle of medicinal spirits concealed in the tackle box** (see footnote). The two men then drape an arm each over the top rail of the stall, take a couple of medicinal pulls from the bottle, and commence to discuss the operation, mentioning the calf is "pretty big for a Caesarean," and comment that there seems to be more Caesarians this year than ever before.

Eventually, you observe them advancing toward the house. Have coffee and cookies ready. As you pour the coffee, add a short splash of household medicinal bourbon to each cup. Make this a very obvious gesture. Place the bottle in the center of the table next to the cookie plate. Then stand back and watch the peculiar glances the men exchange—which could be described as sheepish.

*Old, experienced cowmen always break in young, aspiring cowhands by telling them their job is to "lick off the green calves."

**In certain situations, a Country Man exhibits amazing forethought.

VETERINARIANS ARE MY WEAKNESS

(Tune: Battle Hymn of the Republic)

Mine eyes have seen the glory of the coming of the Vet
He is trampling in the barnyard where the creatures he has met
Have loosed their wrath upon him and this is what he'll get—
Some poop upon his boots.

Chorus—
Glory, glory Veterinarian
Glory, glory Veterinarian
Glory, glory Veterinarian
With poop upon his boots.

He wrassles snorty critters who are sometimes on the fight
He has seen them in the morning and he's seen them in the night
He doctors all the rank ones and he tries with all his might
Not to step in any poop.
Chorus

A farmer had a critter that was doin' bad somehow
The Vet said bring a sample in of fecal from that cow
I'll check it for some parasites and remedy things now
With poop upon my boots.
Chorus

The farmer looked real puzzled and he sadly shook his head
My cow don't have no fecal Doc, cuz she is mostly red
So go ahead and fix her Doc afore she turns up dead
With poop upon her snoot.
Chorus

A horse was full of colic, and his belly hurt him so
Doc hauled him to the clinic where the people usually go

And zapped him with an X-ray quick before the nurse should know
With poop upon his boots.
Chorus

He gave a bull some Rompun and he gave him Lidocaine
The bull fell in a coma and for weeks he felt no pain
He lay around real quiet and that bull then went insane
And soaked poor Doc with poop.
Chorus

He freezes in the winter and he swelters in the heat
He tackles all the problems and he never does retreat
Oh, he's swift to preg-test bovines and he's quick upon his feet
When standing in the chute.
Chorus

In the beauty of the cowpies, the Vet is there for you and me
With his vaccines and his pullers and his A-I plastic sleeve
He will tend our sickly critters, but he will not do it free
With poop upon his boots.
Chorus

COW CLEANLINESS

SECTION SIXTEEN
Garden Maintenance

Wherein one learns to name weeds by their Anglo-Saxon subtitles. One also learns the true aggressive nature of peas, beans, and all those other garden vegetables. Advice offered on how to Tom Sawyer the cucumbers and zucchini.

A major part of summer is spent grubbing about in THE GARDEN. Watering, irrigating, weeding, and hoeing are words used to describe a sore back, blistered fingers, mud-caked clothes, and a feeling of black panic when it all begins to get away from you.

By the end of summer, your entire body falls easily into the hoe-handle stance. The other favored position is the hands and knees crawl. The crawl position saves your back and allows you to notice from close up which plants have bugs. A drawback is the discoloration of knees. As you crawl, the moist earth grinds through the layers of coverall and blue jean to press into the pores of the kneecaps. Your knees develop a mottled, slightly diseased appearance. It doesn't matter much until, wearing a short dress, you have to sit in the front row at Woman's Club.

Try to maintain a sense of humor when leafing through the gardening magazines and seed catalogs* (see footnote). Those magazines always feature elegant, spiffy ladies dressed in white-denim Italian-tailored outfits with coiffured hair peeking out from under expensive garden chapeaus. Grandly they stand there, holding gleaming, clean garden trowels, while wearing cute little garden gloves untainted by any specks of soil.

The garden picture you present contains some variations from the ideal. However, do not be dismayed. Garden costuming is important to summertime morale. Each summer, buy a new, dime-store, polka-dotted or straw hat with a chiffon ribbon. Approach the garden wearing your new hat; striped bib overalls; your old, red-rubber, high-school rain boots; and old, stained, thin-leather gloves. Except for the mud, the freckles, the broad beam, and the can of mosquito spray clutched to your person, you, too, could pose for an ad. Mosquito spray or lotion is an absolute must. As your garden grows, so do the mosquitoes. Always wipe, spray, or douse with repellent any part of you that is exposed before charging the garden. Early morning and late afternoon hours suit your schedule best for gardening. The mosquitoes know this. They train and sharpen their stingers during midday.

*Leafing through magazines is done while performing one's ablutions. Knowledge of current affairs would be unavailable to ranchers and farmers if it weren't for bathroom libraries.

Should you run out of repellent, used motor oil is an excellent substitute* (see footnote). Both the oil and the repellent stink, but should you be asked to pose for a magazine, the smell won't print.

Plan to weed daily. Plan for your sons to weed daily. Press any passing neighbor kids into service as well. Use force, threats, blackmail, bribery, wheedling, coaxing, cajoling, or any other ploy you can think of to get the weeding done. A constant supply of cookies, Kool-Aid, and iced tea should be available to the weeders.

Chickens and geese can be drafted to help kill weeds and bugs. (However, it is not necessary to reward them with cookies or Kool-Aid.) Banty chickens are especially good at eating slugs. Show a Banty a slug and general excitement ripples through his or her little body. Soon the whole gang zips here and there gobbling up the yummies. Geese eat quack grass and any kind of weed. Take care they don't start on the lettuce leaves, however. It's best to let the geese help garden only when you're there to oversee. Geese automatically fertilize as they garden—about every five seconds. When bored with eating, they tend to congregate at the front door where they fertilize the front steps a lot. At all times be aware of the location of the geese. Encourage them to work at the opposite end of the garden from you because geese are basically sneaks. Geese attack anything or anybody from behind. Your standard garden posture offers the foul birds an easy target.

As your garden grows and grows, it becomes evident you are going to have to DO something with all that produce. Some items like lettuce, spinach, or radishes bear early. Eat a lot and give a whole lot away. Take basketsful to the old-folks home and the Senior Citizen Center. Freeze all the spinach and Swiss chard you can. It tastes good in cold weather, is full of healthful iron, and helps maintain regularity.

As the garden ripens more or less all at once, you can stand in the middle and hear the plants shouting and pleading in a frenzy of emotion to be HARVESTED. Apples ripen, corn is ready, and berries are falling off bushes. There's a distinct feeling the garden is uprising to come after you. Do not panic or fall into a fit.

At the edge of the garden, station an old table, or a bench, or a plank on two sawhorses. On top, keep two or three colanders handy, covered with plastic for cleanliness. Have the garden hose available. As you pick the

*Ordinarily the motor oil is used on the pigs from a contraption known as a pig-oiler.

produce, wash and rinse it right there on the table. Place the washed stuff in the colanders and take to the house. On the stove, a giant pot of water is boiling. (All ranch and farm women learn that "keep the pot boiling" is not merely a useless old saying.) On the back burner place a pot of water—fresh daily. Whenever you head for the garden flip on the burner. When you come in, the water is hot and ready to go. Blanch the vegetables you've brought in with you, drain, stuff into plastic bags, and drop in freezer. Pick, blanch, and freeze as fast as you can because canning season is racing toward you. If you are the particular type, follow any cookbook for length of blanching time. Otherwise, merely slosh the leafy stuff till it wilts a bit, and boil the hard, poddy stuff till it turns bright green.

As cucumbers ripen, they go crazy. Eat as many as possible and make pickles from the rest. If yours is a pickle-loving family, you may not skip pickle making. The thing to do is to find several town ladies who may not have a garden but do enjoy making pickles. Offer these talented women unlimited quantities of cukes in exchange for a jar or two of pickles or relish or whatever they fancy to make. If the cucumber crop doesn't freeze out too early, you can acquire enough pickles and relish to last the family all winter.

An overabundance of zucchini is always usual. You can make bread, you can sauté slices for a veggie dish, you can give as much as possible away to friends and relatives and passing strangers. When all else fails to lessen the zucchini supply, that's when you really appreciate pigs. Number One son has taught Peaches, the sow pig, how to catch a tossed zucchini.

She started with one zucchini
Which expanded until she was screaming
She boiled them, she fried them,
She mixed them with gin
And chug-a-lugged a zucchini Martini

SECTION SEVENTEEN
Varmint Control

Wherein one learns that the beasts of nature are not on your side. One also learns how to get rid of magpies, 'coons, coyotes, foxes, and egg-sucking cats.

On every ranch or farm, hazards, tragedies, and misfortunes lurk in the form of drought, flood, freezing, rustlers, and, particularly, varmints. As a true Country Woman, you develop a certain expertise in varmint control, which means you learn to set out traps, bait, and how to shoot guns.

For every domestic critter, there's a whole gamut of wild varmints and beasts ready to do him in. The coyotes and foxes eat the young lambs and then join the skunks to eat the chickens and geese. The raccoons kill the cats and eat grain from feed troughs, from the grain barrel in the barn, or from the cobs in the garden. Mice, of course, eat everything, including the fleece lining of your saddle. And now we are blessed with wolves who, the government claims, eat only field mice* (see footnote).

If none of the above costs you a year's profit, the town dogs band together and go on a killing rampage slaughtering baby critters in berserk frenzy. If you survive all the usual pitfalls due you for that year, rustlers decide to relieve you of some of the burden of caring for animals** (see footnote).

To combat the ravage of preying varmints, whether they are four- or two-legged, the ranch or farm develops its own troops and armaments. Be fearless. Attend to varmint removal promptly!

To keep the mice population semi-controlled, every farm and ranch have many cats. You, as a thinking Country Woman, keep at least two housecats of the female sex. (Like lions, it's the female who's better at hunting for food). Once she' proved herself a good mouser, get her spayed. She'll stay home and skulk around mouse houses*** (see footnote).

Your man deals with coyotes and foxes by setting out traps or poison depending on what the government is allowing that month. When coyotes are in ascendancy, you may start the season with sixty ewes, each with a frisky lamb or two by her side. As ravaging coyotes and mice-eating wolves lunch on the babies, you stand helpless as the death count goes from ten to twenty to thirty to forty lambs. Profit is gone, meat is gone, wool is gone, but

*In the last few years, there have been a lot of calves, lambs, and foals masquerading as mice.

**Rustlers either remove the entire animal or butcher it immediately, leaving only entrails and tire tracks to mark the spot.

***If you're scared of some dark place, like the root cellar, take a cat with you. Throw her in first with instructions to search and destroy.

the coyotes and wolves are healthily producing litters of eight to ten pups a year—on a diet of your livestock.

Raccoons are difficult to eliminate, partly because they're such cute, furry, black-masked animals. Both sons acquire one or two as a pet at least once. The morning comes, however, when you step into the garden and find all the ears pulled from the cornstalks and the tomatoes ravaged. Harden your heart. It's you or the 'coons. You must permanently remove those crop-destroying varmints. Be brave. (That means pleading with your husband to shoot, trap, poison, or imprison the creatures.) Try leaving a good dog in the garden at night with instructions to scare away raccoons. This is chancy because the canine may abandon his post or sleep through the party. A raccoon can be discouraged from eating ears of corn by putting red pepper into the top of each ear. If you've got half an acre or whole fields planted in corn, you're going to need a lot of pepper and a lot of time and a lot of nose tissue while you work. This is an excellent theory (they say) and works well until it rains.

Magpies are beautiful birds that abound. Handsome, sturdy creatures, they sport a white breast, black head, and tail feathers with gleaming, blue-green highlights. A magpie is also a thief, scavenger, and cannibal. Magpies dive-bomb the song birds off their nests and eat their eggs. They also eat grain, dog food, and your rose hips left out to dry. They fly boldly into the chicken coop, spear an egg, and fly off, egg and all.

Liquidate these winged wonders with a .22 or a shotgun. Your oldest son is happy to sharpen his marksmanship on magpies. You, however, shoot badly and don't like to anyway. But there are days when you would personally strangle any magpie you could catch. Be of good cheer. Your chance is coming.

On a warm, drowsy afternoon when the chickens are all out of the coop, you notice a parade of magpies walking like black-suited politicians into the chicken house. Naturally, you are preparing to leave for town dressed in your best hot-pink pants suit. But you do not hesitate. Instantly reach for that old, hard-wood axe handle you keep behind the woodbox. Carrying the handle, keep cautiously out of the line of sight and tiptoe up to the chicken house.

Carefully open the door, slip through, and shove down the panel over the chicken exit hole in the bottom of the door. Look about and note magpies clinging to the upper corners of the coop, staring at you and growling* (see footnote). Growl back and wave your club. As the birds hysterically swoop past your head, flail about with it (the club, not your head). WARNING: When a particularly aggressive bird flies straight toward you, be careful. A mighty swing can miss the bird and clonk excruciatingly on your own noggin. Your presence is required at the bake sale as coffee-pourer in town. When you show up late, sporting a huge purple protuberance over one eye, no one will swallow the story about magpies being responsible.

You may or may not manage to down all the magpies trapped in the coop. As their numbers dwindle, the last one or two seem to get the message. They cling to the top corners and refuse to fly within clouting range. However, it won't hurt to let a couple of them escape. The survivors apparently carry the word back to their relatives. For quite awhile, magpies steer clear of the henhouse.

Skunks present an overwhelmingly pungent problem. In spring, the dogs and sometimes the horses acquire a delicate perfume. If at all possible, plead with your man and son to deal with skunks. It's best to trap them alive. Your husband has developed a skunk-catching box that has the door opening plus one small, round hole in the roof of the box. After capture, he inserts a piece of hose from the skunk box to the exhaust pipe of the pickup. Then he starts the motor. Exercise selectivity and judgment in attending to skunk removal and burial. Many ranches and farms have small, well-removed gullies or coulees named "Skunk Hollow."

Porcupines can present a prickly problem. If possible, bait them into a box and relocate them miles away. Should you be faced with the necessity to kill a porcupine, don't. Call your husband and refuse to think further about the problem.

Occasionally an outlaw cat becomes hooked on raw eggs. Craftily, the feline felon sneaks into the chicken house, punctures an egg, and laps up the goodies. It's easy to tell which cat is the culprit. He's the one with the lustrous sheen to his coat who sits in the sunny spot in front of the chicken house licking himself and smiling. If Barney, the egg-sucking cat, is tame, round him up and give him to a favorite relative as a loving companion. If Barney is wild, it may become advisable to execute the pest. Again prevail upon your husband** (see footnote).

*Pay no attention to persons who claim birds don't growl.

**Since your husband hates to have to kill anything almost as much as you do, super-rich chocolate cake is the order of the day at these times.

As with other of the basic raw facts of nature, it is not necessary to dwell on varmints and especially their executions in conversation, particularly around town folk or environmental zealots. Should you, for some reason, find yourself harassed by such a person, send the person a gift of a bird cage with a nice, live magpie inside.

"Quit stealing my eggs, you magpie,"
Said the farm gal with blood in her eye.
"You can't catch me,"
Said the bird in great glee.
So she shot him rather than try.

The prairie dog pops from his burrow
And sights a convenient furrow;
It's green and it's sprouting,
So he plans an outing—
At chewing it up, he'll be thorough.

Gophers look cutsie and charming
But some of their ways are alarming.
For one, they'll devour
Each leaf, stalk, and flower
Of whatever crop you are farming.

SECTION EIGHTEEN
Summer Company

Wherein one learns that otherwise polite friends who live in the City tend to look upon you, the Country Woman, as recreation leader for THEIR vacations. One learns some methods of ridding oneself of unwelcome visitors.

Summer company differs from standard drop-in visits of neighbors and friends. Neighbor folks "stop by" on a particular errand or to visit about interesting news such as the weather, the crops, the twin calves, or the upcoming Community Hall Dance.

Summer company consists of persons of slight acquaintance and distant relatives so far removed, intermarriage would not be a problem. Somehow they've developed a sudden close, warm, personal yearning to be near.

As a good Country Woman you will often devote summer Sundays to the care and feeding of company from the City. When the Sundays extend to a week or more, you will plumb your depths for patience and tolerance. This saintly approach will mainly drive you plumb dippy before summer is over.

Generally, summer company arrives in a family package: a man, a wife, from two to six children, and a pet dog. (Always a big German shepherd or a neurotic toy canine of some sort, or both.) None of the group knows anything about the country. But my, oh my, are they enthusiastic!

"So wonderful to breathe the fresh country air," they burble, as the dog roars out of the car and rips off a chicken. (You, naturally, assure them that it's all right—it was just an old rooster—as your best laying hen gasps out her last.) The visitors seem convinced the entire countryside has been unrolled for their special pleasure during weekends and vacations. After that it's all rolled up again and put in storage till they come again next year.

In summer, you, as the good Country Woman, will struggle out of bed anywhere between five and six a.m. (or earlier). That allows plenty of time to prepare breakfast for your husband, family, and the hired hands. Then trot out to feed the chickens, the pigs, and the bum lambs before trotting back to the house to fix breakfast for THE COMPANY. The company pull themselves out of the nest from eightish up to noonish.

Country custom says your job is to be cheerfully chatty as you serve up platters of "real country breakfast." That means lots of eggs, bacon or sausage, and hotcakes with two kinds of homemade syrup. Smile when Mrs.

Mother Company has to cut ten-year-old Junior's bacon and egg for him. Smile when nine-year-old Juniorette loftily refuses her portion of breakfast and demands cornflakes. Continue smiling when the dear child sprinkles out twelve flakes into a dish, smothers them under a half cup of sugar, then drowns everything under a flood of cream. Smile, when after two bites, Juniorette decides she doesn't like all that cream, takes five cookies from the jar, and exits, leaving a globby mess strewn on the kitchen table* (see footnote). Keep smiling, but move swiftly to intercept Junior as he leaves toting the bacon platter. His intentions are to feed the remaining eight pieces to King-the-Collie. (You had planned to dice up the leftover bacon into the scalloped corn casserole for dinner.)

Over the years, summer company falls into recognizable types.

Type A: That would be the loud-talking family group who never lets you know ahead of time that they're coming. They pull into the yard unannounced always at chore time. They've been "sight-seeing" in the nearest city or national park a hundred miles away for three days, but hadn't thought to call ahead. Naturally they expect to stay for supper and overnight lodging. (You pray it's ONLY overnight.)

Type B: The barely acquainted or never-heard-of persons referred to you by mutual "friends." The "friends" are generally old army buddies of your husband's or former coworkers from that summer your man worked in the oil fields or a national park before you were married. Or they know his or your sister in Peoria. Country custom says you must find beds, sleeping bags (you hope the little kids in bags won't wet), and floor space for all** (see footnote).

As the company start up the walk, throw your dish towel over one shoulder, go to the door, and welcome them cheerfully. Mentally calculate the increase in hungry mouths and break out the card tables for the kids to eat from. Sit up late that night exchanging conversational trivia with the company. Find out early whether the visitors plan to leave next day or are preparing to camp with you "for a few days." If it begins to look like you may have the pleasure of their presence for very long, start lacing the conversation with comments about how bad the rattlesnakes are this year. Mention your near heart attack when you reached for the mop bucket and

*King-the-Collie loves summer company. Into his dish go all the yummy cream leftovers and half-eaten eggs. In addition, he usually manages to steal at least one cookie per kid handful.

**Ranch and farm houses are expected to come equipped with fold-out, fold-up, bunk, and trundle beds tucked away in various crannies.

disturbed a big snake at least eight feet long. Describe how it hissed and struck at you, narrowly missing your hand. If young children are among the group, mention an outbreak of a new strain of measles in the area. Add that your son has been exposed and you're a little worried because there's no vaccine yet developed for this new measles bug.

Type C: This company includes relatives close and relatives distant. Close relatives, such as in-laws, are inclined to snub the already entrenched company. Conversation at the dinner table drags to a halt as you and your husband try desperately to introduce a subject of interest to all. Immediately after the meal, your man disappears to work on the tractor, doctor the calves, or check the windmill. For whatever reason, he fades away leaving you with two groups of persons talking at you, but not to each other. As soon as the dishes are finished (by you, while the company watches) put on your coveralls, grab a shovel, and invite everybody to help clean the chicken house. Whether anyone helps or not is immaterial. By the time you get back to the house, at least one of the company groups will have decided to move on.

Type D: This company includes those relatives or friends who are rich and/or imposing. Your husband's Aunt Bertha, whom he hasn't seen in ten years, always sends the kids expensive Christmas presents accompanied by long letters of helpful child-raising advice. On the worst possible day, Auntie Bertha phones from the local country store for proper directions to the ranch. Obviously there's only one thing to do, short of running away from home. Instruct her carefully in ALMOST the right direction. With only a very small slip of the tongue, you can steer her left instead of right at the crossroads. By the time she stops and asks and backtracks, you will have gained perhaps a whole hour to GET READY.

Last week's company, including four kids, just left and you haven't had time to do any picking up. Therefore, the house is carpeted wall to wall with a mixture of toys, rock collections, picture books, clothes, and a chaotic jumble of unnamed THINGS. Do not panic. Take a deep breath and scream once to clear your head. Then concentrate on scooping up the clutter and stuffing it into closets. Hastily run the vacuum around the middle of the rooms. Quickly swish-wipe the bathroom and dust the guestroom. Open the package of brand new sheets kept for emergencies only and apply to guest bed. Stack all the dirty dishes in the dishwasher. What's left over, put in brown paper sacks and shove under the sink. (That night after Auntie retires, fish them out and wash.) Race to the garden and pick some sort of flowers. Jam them in the crystal vase you're pretty sure Aunt Bertha gave you several Christmases ago, and place it on the dresser in the guestroom.

Just as you finish the fastest cleanup in the West, Aunt Bertha drives into the yard, where she is first greeted by four big, white Canada geese, honking and hissing. Arrange your smile, carry your dishtowel, and go out to welcome Aunt Bertha. IMPORTANT: Carrying the dishtowel is an absolute must. It is a dual-purpose implement. Besides giving you something to wring other than your hands, it can be employed to discipline a friendly seventy-pound collie and fend off honking, hissing, wing-flapping geese. Aunt Bertha cringes before the snapping onslaught, whereupon the geese press the attack and bite her nyloned legs* (see footnote).

As the summer progresses, the guest traffic increases, particularly if you are on the route to a national park, the World's Fair, or own a saddle horse. All city people "rode a horse when they were little." Such past expertise equips them to "round up the cattle." Whether the cattle should be rounded up has no bearing. The company wants to do something "typically country." This includes riding, roping, shooting, and eating. It excludes hay-stacking, irrigating, and manure shoveling.

Common to most types of summer company is an obsessive urge to PLAN. They'll try to fit into two weeks every scenic wonder possible. You are drafted as the local guide and social leader to "show them the sights." Which means you must plan two days' meals for your family in advance. Leave the butter churning till you return, along with the separator washing, the garden watering, weed-pulling, and canning and freezing chores. If you mention these duties in a desperate effort to stave off the inevitable, the company gaily dismisses them all and urges you to "take a day off and relax." They claim they will help you upon returning. (They won't. They're always too tired.)

However, give in gracefully, because your husband is also craftily urging you to go. Even though there will be extra chores for him, he would rather see you herd the guests out of his way. This allows him time to get his farming, or haying, or some other major project underway without interruption from "helpful" company.

Helpful company can drive a good man batty. For three days, your husband has rescued greenhorns from being mashed by a tractor, trampled by

*Don't forget upon entering the house, with Aunt Bertha in tow, to discard the now goose-soiled dish towel. Aunt will remind you if you forget. Make an obvious show of selecting a fresh, clean towel.

cows, and drowned in the irrigation ditch. One entire evening was spent searching for the company's German shepherd who had gotten himself lost while following his horseback-riding master. (Mounted on Old Dobbin, a bomb-proof gelding. Give Old Dobbin extra oats if he brings the kid back safely.)

Be a good Country Wife and take the guests off on a trip, leaving only their Cousin Agnes behind (who sprained her ankle yesterday when she fell off the corral fence). Upon your return, you will find that Cousin Agnes has helpfully cleaned all your kitchen cupboards. She has also efficiently reorganized your entire storage scheme. It's been a year now and you still haven't found the nutmeg.

As summer draws to a close, you think you can begin to relax. School opens in a week. After that, people have to stay home with their kids, don't they? Be thankful for compulsory schooling. Enjoy your few moments of solitude. Just around the corner is hunting season when hordes of hunters will be turned loose upon the land—generally, yours.

SECTION NINETEEN

Canning, Freezing, and "Putting Up" All that Garden Produce

Wherein one learns that Mother Nature can overwhelm. One also learns that frontier life without electricity and freezers must have been doggoned challenging.

To be a Ranch or Farm Wife is to be afflicted with an overpowering urge to can. NOT to can all that garden produce would cause an unbearable burden of guilt. Therefore, beginning in midsummer, you, as a True Blue Country Woman, must drag out the canning kettle, the pressure cooker, and a million jars.

Even thinking about canning causes a sticky-footed lurch to your step as you remember the gluey, syrupy mess from spilled-over and boiled-over liquids. The easiest way of preserving food is to freeze it. Anything will freeze, but sometimes the flavor or texture is changed in the process. Frozen lemon pie tends to taste like frozen sawdust. Generally, a working combination of canning and freezing is employed when preserving food.

Lurking in the storage shed and fruit cellar are baskets and boxes of empty fruit jars, jelly jars, odd bottles, and interesting glass containers* (see footnote). Clustered in the top kitchen cupboards—never used ordinarily, because they are too high to reach without a ladder—are more bottles and jars of assorted sizes and shapes. No Country Woman has enough jars. Each year the need for more increases until your husband becomes convinced you're a jar addict. Jar collecting is, however, merely a desperate and futile hedge against the avalanche of fruits and vegetables that must be "put up." Somehow you have a feeling that if you collect enough jars, canning won't be so bad this year. But it will be. Canning will be bad this year and next year and on and on forever. But some of the rewards are being able to brag to the neighbor country women how many jillions of jars you've "put up."

There are two periods of time within the twenty-four hour cycle that are best for canning. At these times, other humans and all animals are quietly asleep. Only you are astir, cackling softly over your brews. Early morning is the favored time choice for many, oh very early. Stagger out of

*The fruit cellar, over the winter months, has become a hatchery for creepy wildlife. Spiders and bugs dangle from the roof beams, and you're positive an unseen monster is skulking in the black hole behind the potatoes. It's comforting to take both the cat and the dog along. At least they'll raise an alarm if something grabs you.

bed by 4 a.m. at least. The other time slot preferred by some Country Women is the ten p.m. to one a.m. shift* (see footnote).

Whichever time you choose, remember to PREPARE ahead. Clear the kitchen table, the countertops, and every surface in sight of the usual litter. Place canning kettles or pressure cookers on stove. Have plenty of long-handled wooden spoons at the ready. Line up a stack of boxes of fruit jars on standby alert on the back porch. On the sink counter, have washed, scalded, and turned upside down on clean cloths a multitude of jars. Be sure you have plenty of lids of the correct size!

There is no way to make canning really fun, unless you hire someone to do it for you. The canning season usually begins with berries. The juice has to be extracted and canned and later made into jellies and syrups. Having picked gallons and gallons of whatever berry is presently ripe, dump them in the biggest kettle, cover with water, and heat to boiling. If you are fussy, carefully wash berries and pick out extraneous leaves and bugs before heating. However, a few stirs after the kettle begins to heat will bring all that flotsam floating to the top. Then you can skim most of it off. What's left won't hurt the flavor and you're going to strain it all anyway.

For just how long to cook what and which method to use, consult any good canning and preserving book. It won't be long before you will have every step memorized, as you can and can and can day after day after day ad nauseam.

Sometimes when you've nobly risen at 4 a.m. to fight the canning battle, you find yourself having visions of South Sea islands where food drops off trees. Or you find yourself making up limericks having to do with banning canning. Or stray lines from the dim past when you read Shakespeare flicker in your mind. As you swirl your wooden spoon among the vapor clouds emanating from the giant kettle boiling on the stove, you find yourself muttering something about "bubble, bubble, toil and trouble." Generally, just at the moment when you're really getting into the swing of the recitation, complete with eerie cackling, your man stumbles into the kitchen for his morning cup of coffee. Peering at you through the mists, he lifts his cup to misquote solemnly, "a rag, a bone, a hank o' hair; by God, there stands my lady fair." It's best to accept this bit of humor in the vein in which it is offered. Burning your finger on the kettle and bursting into tears helps also.

Early morning canning sessions aren't too bad once you've conquered

*Be careful not to sit down in a soft chair during the evening shift. To do so is to fall asleep. At which point whatever is cooking immediately boils over.

the pain of getting up and getting started. Late-night sessions, however, can be more exhausting. As the hour grows later, the mountain of food to be canned seems to increase. Your feet hurt; your back aches. As you stare at the bubbling cauldron, you wish to God the entire outfit would blow itself up. Somehow you feel that would put an end to the whole canning question.

During the late-night sessions, as the hour grows later and tireder, it is recommended to take two aspirins and a glass of soda pop. In the bottom of the glass, first pour in a heaping shot of bourbon. If the pain in your back, legs, and feet does not diminish within fifteen minutes, add more bourbon* (see footnote).

A pardonable habit of all Ranch and Farm Women is to march any visitors down to the basement or out to the root cellar to show them the rows and rows and rows of shining jars of canned tomatoes, green and yellow vegetables, fruits, jellies, syrups, meats, and pickles. Each jar is neatly labeled with its year of birth. You are filled with satisfaction and pride as you view the display. Insist, therefore, on hearing admiring oh's and ah's from your visitors. Mention how much you will be saving on the food bill and how good home-canned food tastes as compared with that plastic store stuff.

Often, it is well into fall before you can bear to disturb those rows of sparkling jars. Once an inroad has been made, it's not long before there's practically nothing left and you are again facing armies of empty jars waiting to be filled.

The Country Woman got out her jars
And peeled and pickled for hours
She got so darned tired
She nearly expired
But revived after nine whiskey sours.

*Exercise caution with this remedy. The aim is to blot out your physical pain without dissolving yourself.

CHOKECHERRY SYRUP AND JELLY

You can't beat homemade syrups and jellies. Chokecherries are abundant most everywhere. And they're fun to pick and do up beautifully. On a fine morning, garb yourself in your usual jeans but add a belt or a bungee cord around the waist. That's so you can hang the five-gallon, empty, ice-cream bucket from your belt, leaving both hands free for picking. Load the pickup with a big cardboard box or bushel basket lined with a plastic sack; also, a ladder to conquer tall bushes. Drive to your favorite lane or road where the chokecherry bushes and trees await. When you fill the bucket at your waist with the juicy orbs, dump the contents into the box or basket. You don't want to overfill the large containers or you won't be able to lift them, so take along extra plastic replacement sacks.

TO OBTAIN JUICE

Wash the berries some and put in large kettle with water to cover (about an inch over). Bring to a fast boil. Reduce to simmer and simmer one hour. Remove from heat and drain through colander (don't mash). Then strain that liquid through cheesecloth. You can then can the juice for later use or go ahead and make syrup and jelly right away.

CHOKECHERRY SYRUP

Use unsweetened apple juice with choke juice. The apple juice cuts a little of the tangy, pungent edge of the chokecherries without changing the flavor. With every two cups of choke juice, use one cup apple juice.

In a large kettle, put 4 cups chokecherry juice and 2 cups apple juice. Add 1 package powdered pectin and bring to a full rolling boil. Add 6 cups of sugar all at once. Then boil 4 minutes. Remove, skim, cool 5 minutes, skim, and bottle. (Depending on where you live, altitude-wise, you may boil a longer or shorter time. Experiment. Your kitchen is your laboratory.)

CHOKECHERRY JELLY

In a big kettle, combine:

3 cups choke juice

1 1/2 cups apple juice

1/2 cup lemon juice

1 package powdered pectin

Stir and bring to full, rolling boil. Add 6 cups sugar all at once. Bring to rolling boil again and boil 5 or 6 minutes until jelly sheets from spoon. Skim, cool 5 minutes, skim again, and put into jars. Seal with paraffin.

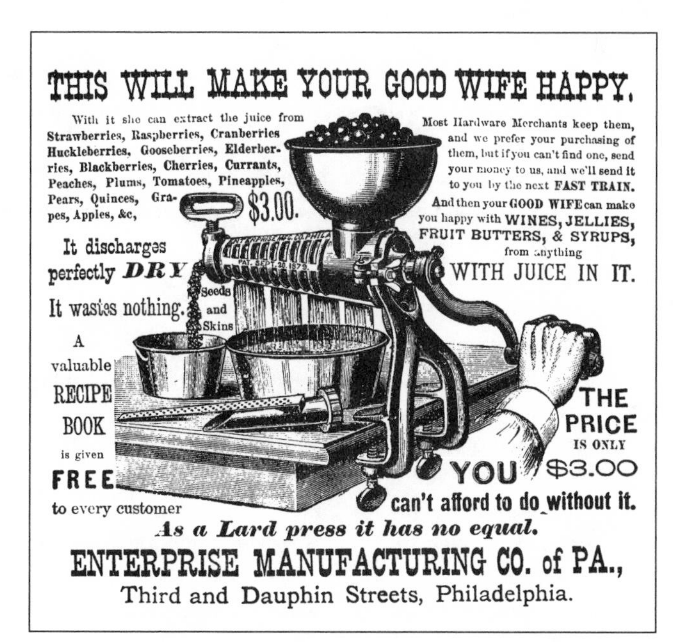

SECTION TWENTY
Summer Fun

Wherein one learns the correct method of country tippling and how to dance country style.

Summer dances in the country are held at Swamp Creek Community Hall. The occasion truly enriches your life while debilitating your system. You drive to the hall in the pickup. Important preparations must be made. A cooler filled with soft drinks, ice cubes, paper cups, and beer is fitted in on your side of the pickup underfoot. The hard liquor is kept in the glove compartment. Make heaps of sandwiches or a giant, heavily frosted cake to carry into the dance hall and place with other donations in the hall kitchen. At midnight, you and the other Country Women will lay out the food and pour the coffee—just like home.

The cooler with the beer and the glove-compartment beverages remain AT ALL TIMES in the pickup* (see footnote).

In the hall itself, the men congregate in a bottleneck at the front door. Wooden benches lining the hall support long rows of women and children. You sit on a bench and "gab" with the other women.

From time to time, small groups of men quietly leave the hall and casually head for the vehicles. Later on, the departures and reentries become noisier while the dancing markedly increases. No realistic *married* Country Woman can expect actually to dance for the first two hours.

You are introduced from time to time to other ranchers and farmers and their wives. Not necessarily by name, however. "Like you to meet the wife," or, "Have you met my woman?" is about as close to hearing your actual name uttered as you can expect.

At midnight, a line of hungry noisy revelers clomp past the food table, where a large commandeered cowboy hat is held by a large commandeered cowboy who collects money from the men to pay the musicians (always two guitars and an accordion and, maybe, a fiddle).

After food intake, the Country Men settle down to serious drinking and dancing. Country custom says that, when asked, you must dance every dance as long as your strength and feet hold out. Try not to shudder at the piercing war whoops (especially from the single young buckaroos looking to impress the girls), which punctuate the end of every number.

*With the onset of Big Brother laws which proclaim restrictions are for our "protection," it may be necessary to assume a saintly mien and keep the stash corked and under the seat.

Since midnight you have refrained from drinking hard liquor. (Sometimes you accompany your man to the delights of glove-compartment indulgence. If the bottle is passed to you, only pretend to swallow since at four or five in the morning, SOMEBODY has to be sober enough to drive home.) As a true Country Woman, you're a proficient driver of any style vehicle, especially at those times when you are the Designated Driver.

Always, on the eve of the Annual Rodeo, most of the populace of any country town and county environs turn out to dance, imbibe, and catch up on local news. Every bar and every fraternal organization holds a "dance" with "live" music. Which means anybody who can play at least three tunes on a guitar qualifies as a musician. Drinks are served in paper, Styrofoam, or plastic cups at all the liquor-dispensing emporiums. This convenience allows imbibers to port their drinks into and out of bars, across streets, and up and down sidewalks. The imitation glassware does not break, thus saving feet and fingers from possible laceration. Bar operators do not have to replace smashed or lost glass inventory and the town doctor saves on bandages.

In country towns, no one arrives at a dance before ten p.m. in the summer. Animals must be tended and chores must be done first. As dusk settles in, your husband "gets ready to go to town." His going-to-town clothes consist of clean Levis, shirt, no tie, his "dress" hat and his fancy pair of cowboy boots. Your costume is much the same except you don't have to wear a hat unless you're competing in the Rodeo Queen contest. Or you may yield to a certain nostalgic yearning and wear a dress. Whatever your choice, always remember to wear low-heeled, comfortable shoes or boots.

During much of the evening, you stand around holding a drink and watching all the other people. No ranch or farm man ever sits down at a public dance unless he's in his dotage. Along with other Levi-garbed males, your man bottlenecks in the doorways to watch other folks, listen to the musicians, and "visit" with neighbors.

As the evening progresses and your feet play out, you learn to develop restful leaning methods. Lean for awhile against your husband if he's nearby. Then lean for awhile on the juke box, the piano (if there is one), or the pool table (most likely to be one). If you get near the bar, lean on it. When the eddies of people carry you near the edge of the room, it's quite restful to double one leg and prop oneself, stork like, against a wall. Glance into any country bar late on a festive night and you will see a lot of apparently one-legged women lining the walls. While this position can be restful, take care lest the doubled-up leg goes to sleep causing a possible collapse as you try to walk.

Some years the town fathers and bar owners decide to have a street dance. Down the center of the block of Main Street, sawhorse tables blossom. Each end of the street is closed to car traffic. Bar owners haul out basic beverage supplies and ice in washtubs. At intervals up and down the sidewalks, musicians station themselves to play.

There's something about a street dance that drives otherwise sane and sensible farmers and ranchers slightly goofy. During the festivities it is best that you and your husband develop a strategy. Take standing positions in the shelter of a doorway next to a bar. Usually there's music in the bar as well as on the street. So your chances of getting to dance are increased* (see footnote).

For weeks and months after a street dance, the countryside resounds with wild tales, most of which are true. Legends are born and sometimes politicians elected on the basis of campaign promises on the eve of events like the annual Bull-A-Rama Rodeo.

Such was the case with one of the county commissioners recently reelected. After careful consideration and half a quart of whiskey, the would-be commissioner decided to pass around his campaign cards. Accordingly, he saddled his old, white mare and proceeded to ride her around and around the block-long street bar, passing out cards and handshakes. He also accepted drinks whenever offered, which was not infrequent. The inevitable occurred. It struck the campaigner that the people INSIDE the buildings were being deprived of his cards. Methodically, the aspiring politician guided the mare into every bar and dance hall in town. Solemnly, he proffered cards to people seated in booths, dancing couples, and those going to and fro.

It was while crossing the dance floor at the Fraternal Lodge Hall that the mare became nervous and deposited some business in the middle of the floor. Amid roars and shrieks of laughter, the horseback politician pulled his steed to a halt to view the somewhat steamy pile behind him. As the laughers quieted slightly to regain their breaths, the commissioner rose tall

*Even more important, you have access to a ladies' room. No matter how liberated you are, it is not a good idea to slip down an alley. A not-so-liberated man is apt to be lurking there with a similar chore in mind.

in his stirrups, swept off his hat, and proclaimed, "Folks, if you want your business done right, vote for me!" He was voted into office by an overwhelming majority and has held the post for eight years.

Lucky is the Country Woman whose spouse can and will dance. Likely though, you are among the vast sisterhood of CWY (Country Women Yearners). As the music pours over and around you, the yearning to really dance becomes an actual pain. Every person in the room moves to the music except for the contingent of Country Husbands clumped in the doorway and captained by your spouse.

Generally, you are driven to desperation and have to implore and plead. Surprised at such an odd request, your man dutifully escorts you onto the dance floor. Having reached a location he likes, he commences guiding you about more or less in time with the music. Duty done, he again takes up his vigil near the door, satisfied that you've had your fun.

There are some gems among Country Men who truly like to dance and actually go about inviting the ladies! Be sure to get on that man's list for at least one twirl. That will make two times for the evening, anyway.

As the liquor consumption increases, the clots of men stationed in doorways begin to filter onto the dance floor. Like sleeping bears, the men rouse to the music* (see footnote). For awhile you find yourself whirling and stomping around the room with a variety of partners. You must be brave as you adapt courageously to the varied dancing forms** (see footnote).

Although styles vary, basic country dancing is always the two-step—sometimes slow and sometimes fast. The two steps can be two steps forward and back, or two forward and forward, or backward and backward or from side to side. While two-stepping, you are piloted round the floor or across it. Your feet follow your partner in whatever direction he's going, while your right hand is clasped in his left, in a grip as if he's hanging onto a snaky critter. Those two clasped hands often describe a rhythmic pattern not necessarily related to the music or the two-step. Sometimes your hand and arm is lowered and raised from knee level to shoulder with each drumbeat. Sometimes your partner caroms at a gallop around the floor. You, of course, gallop along, too. Holding your hand in his, in rigid extension, your partner uses the pair of stiff arms to break trail through the thickets of people. Hang on tight to his shoulder with your remaining hand. The galloper is usually

*Imbibing all that firewater helps, too. The trick is to get in as many dances as possible before the men get thirsty again. Or before they become so inebriated, they can't maneuver.
**Those low-heeled comfortable shoes mentioned earlier should also be STURDY!

programmed for one direction only and you're bound to be dizzy before it's all over.

Country custom says you may not refuse a dance request. Besides you wouldn't want to. That surfeit of dancing won't last long before the men disappear to quench a desperate thirst again.

> The countryman who seems to prancing,
> Retreating and later advancing
> As if in some pain;
> Let me explain,
> It's just his idea of dancing.

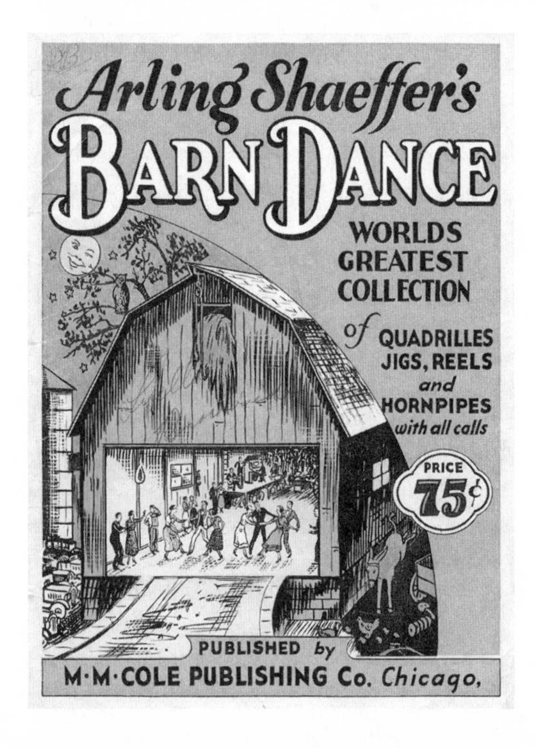

PART THREE

Fall—When The Year Starts Its Downhill Slide

Fall: A gorgeous and beautiful time of year when you can't believe summer is over. The critters grow lots of hair and a great deal of energy is spent discussing MARKET PRICES.

SECTION TWENTY-ONE

Roundup Time and Marketing the Calves

Wherein one again becomes an instant cowpoke. One also learns that the stockyards are a man's world and, after an hour, remarkably boring.

As the ranch or farm swings into the rhythm of the fall season, the animals grow sleek and shiny. Calves begin to change from fractious, funny youngsters to stocky, cud-chewing cattle. A short, but peaceful, lull of good living, good eating, and good health pervades the ranch. Then one day, that smooth ribbon of peace develops a crinkle. MARKET REPORTS blare over radio and TV. Your husband listens intently to fast talking, twangy-accented announcers, auctioneers, and cattle buyers singing their market songs. Dire predications fall from your mate's lips. Gloomily he searches for reassurance but without real hope.

Listen sympathetically as he worries that the calves will be too light to bring a good price. Remain calm as he fidgets waiting to see if prices will go up. Don't argue while he gloomily predicts the bottom will fall out of the market. Cluck compassionately when he says he won't be able to buy feed for the cows and pigs this winter—but it won't matter because he'll have to forfeit the place to pay the bills anyway.

As a good Country Woman, do not mention any activity that costs money. Try not to have dental problems, mechanical failures, or have a baby previous to marketing the calves.

Keep a supply of old envelopes on the kitchen table so he can "figure"* (see footnote).

The morning finally arrives when your man decides it's time to bring in the cows and calves. The kids stay home from school and a couple of neighbors sign on to help out. At an unreal hour in the predawn (about three a.m.) you serve breakfast to the entire bunch.

As a thinking Country Woman, make several pots of coffee the night before and stir up a giant batch of hotcake batter. Country custom says the pancakes should be "sourdough." However, fiddling around with all that may be more than you can stand previous to the drive. Don't say anything. Simply build the pancakes with buttermilk. Make the batter a little thin and put in lots of baking powder, a little sugar, but NO soda. The cakes will be puffy, golden and taste a trifle sour. Drop the word "sourdough" once or twice and all will gulp the cakes with relish. To cinch your reputation as an early morning hostess, add a few blueberries to the batter. Serve the usual vast quantities of eggs and bacon and keep the coffee coming.

Should you have the time and inclination to prepare actual sourdough pancakes, here's a dandy way to make the "starter," which can be kept for years and years and years. Some old-timers trace their sourdough starter back to when the Pilgrims landed and then traveled along the trail as the nation expanded westward. Feel free to make up your own legends.

Whether genuine or pseudo hotcakes, in a flash, the mountain of breakfast is done away with. It's still a long time before daylight. The men can't start to gather the cattle because they can't see. They've gotten up early in order to "get ready" to ride, which means sitting around the breakfast table discussing market prices and drinking coffee (which you pour)** (see footnote).

Eventually, a thin grayish tone sifts into the sky. It's possible to discern faintly the outlines of the garden gate through the window. The men comment upon this phenomenon for awhile. At a secret signal, they all rise, clap hats on heads, and go out to saddle up. Your dear husband saddles your mare for you, thus giving you time to clear away the mess from the table and put last minute things into the oven or refrigerator.

*For a Country Man, the backs of old envelopes are first choice for figuring anything. After that, it's paper napkins, the edges of newspapers, or the cover and margins of the telephone book.

**You, personally, are never actually noticed as a living being. The only reality is the coffee pot attached to the end of your arm.

There's an old tradition in some western circles, which says everyone saddles his own mount. That includes women in particular so it won't appear a man is "waiting on" a woman. If ever your man subscribes to this sentiment, it can leave you frantically clearing up the kitchen, racing for the barn, hurriedly saddling up, and galloping after the men. When you catch up, you're apt to be greeted with clever remarks about being late or the way you bounce when your ride. Clearly, the thinking Country Woman must devise a subtle hint to encourage chivalry. Therefore, the moment the men establish themselves on their feet preparing to head for the corral and barn, quickly don your riding boots, jacket, hat, and gloves* (see footnote). (IMPORTANT: For roundup duty, be sure to wear a sturdy sports bra.)

On your way out the door, stop by the kitchen table. Do not—repeat—do NOT remove a single object! Leave dishes, syrup, leftover eggs, ketchup—everything. Gather up the four corners of the tablecloth. Bring them together and tie a nice knot. Grasp the bundle firmly and tote to the large incinerator which stands on the path between the house and the barn. Drop bundle in incinerator. Calmly continue to barn and saddle your mare. Later that day, let your man discover the deposit. When he asks, "Why??!" (he will certainly do that), mention sweetly that for you it's either clean up or saddle up. (This is a can't-fail recipe.)

An air of tense, frustrated hurry permeates the early morning roundup. The cows and calves have to be corralled. Mammas and offspring have to be separated and the calves trucked to loading corrals outside of town by ten o'clock to meet the BUYER. Separating cows from calves is not always a smooth operation. After the critters are herded together in the big corral, the men deftly edge them a few at a time toward the small corral that is the anteroom to the loading chute. Your job is of prime importance. It's up to you to handle the swinging gate leading into the small corral. As a cow-calf pair heads your way, urged forward by grinning Number One son, you must open the gate at the proper moment, let the calf slip through, and quickly slam the gate closed in the cow's face. Sometimes you miss, and both a mother and calf enter your gate at once. Be prepared with a special curse word at this point. If you do not feel comfortable using hard-core swear words, develop a repertoire of substitutes in French or Russian. You will need something vehement when a heavy pole gate reverberates off a thousand pound cow to slam back on you. As a satisfactory expletive, "naughty, naughty" just won't do.

*It's best to line up your own and the kids' riding outfits the night before. Boots, jackets, hats, and arms and legs can be difficult to find in the early morning.

SOURDOUGH STARTER AND PANCAKES

STARTER
1/2 cake compressed yeast *or* 1/2 package active dry yeast
2 1/2 cups lukewarm water
1 tablespoon sugar
2 cups flour
Soften the yeast in 1/2 cup water. Add the rest of the water, sugar, and flour. Mix well. Let stand in a covered bowl or crock (not metal) for three days at room temperature. STIR DOWN DAILY! Refrigerate after the three days have elapsed.

PANCAKES
When you are ready to make pancakes . . .

The evening before:
Put one cup of the starter into a largish bowl (put the rest back in the refrigerator). Add 2 cups lukewarm water, about 2 1/2 cups flour, and 1 tablespoon sugar. Mix well. The mixture will be thick and lumpy. That's OK! Cover the bowl and set in a warm place (not hot!) overnight.

Next morning:
Take one or two cups of batter from the bowl and *put back into the starter jar in the fridge.* (Each time you set out sourdough batter overnight, you always put back a cup or two into the refrigerator batter the next morning. How did you think the stuff was handed down?!)

Then, to the batter remaining in your pancake bowl, add:
1 egg
2 tablespoons cooking oil
1/4 cup instant milk or evaporated milk or cream
Mix into the batter thoroughly.

Then, *in a separate bowl,* combine:
1 teaspoon salt
1 teaspoon baking soda
1 tablespoon sugar
Mix well, pressing out the lumps of soda to a smooth, dry mix. Fold *gently* into batter in first bowl. This causes a foaming, rising action. Allow to stand for a few minutes, then spoon batter onto hot, greased griddle. If batter seems too thick, add a small amount of cream or liquid milk. *Never add more flour to your fermented refrigerator starter. That's a no-no.*

Sourdough requires a hotter griddle than conventional pancakes. The above recipe serves about four. Hint: Pancake eaters can put away more sourdough cakes than the standard cakes. (Sourdough intake is a pleasant addiction and has no side effects.)

It's always a half hour late when the calves, having been encouraged up the loading chute onto the stock truck, finally reach the rendezvous point. There, the Red-Nosed Buyer has the option to accept or reject any calf* (see footnote). When prices are high and the buyer can turn all the critters he can obtain for a profit, he takes every calf without question. However, when it's a bad year and you desperately need the money, the buyer cuts back small calves, off-color calves, short-tailed calves, long-eared calves, or anything else he imagines doesn't look right.

In prosperous years, after the deal is consummated and the calves are on their way to a feedlot, your husband is likely to generously invite all the men who helped to have dinner and drinks at the local restaurant. (On him, of course!) In that event, you always go along and reign as Queen of the Mayhem. It's a double pleasure because it means you don't have to cook and serve meals to all those people—or clean up after them.

Some years, your husband prefers to haul the calves to market at auction yards in the City. The same routine prevails as regards feeding everybody at breakfast and rounding up in the dawn's early light.

*Cattle buyers come in tall, short, fat, and thin sizes. For some reason most of them have red noses. They are given to a lot of squinting and hand-flapping as they accept or reject each critter.

If calves are going to the auction yards, your routine takes a slight deviation. As soon as the calves are sorted, abandon your gate duty and race for the house. You have previously organized your "town" clothes. Quickly change into that dressy pants suit; polished, sand-colored boots; and smart, suede leather coat your husband bought you for Christmas* (see footnote). Add small earrings (if you're the earring type); your posh, fawn-colored Western hat; and don't forget mascara, for the long-lashed look. In the City, putting on makeup comes first. However, in the Country, it's best to leave personal face beautification until last. Should you be required to hit the trail immediately—if not sooner—at least you'll be attired attractively. Then, you can do your face in the vehicle while traveling. All Country Women learn to apply lipstick while riding in a two-ton truck over bumpy roads.

Eighty miles down the road, the caravan of vehicles full of calves arrives at the auction yards. After unloading, the men step into the "chute office." Country custom says you do not enter the office with them, but hang around outside. Pay no attention to custom. Go on in. All over the office walls are various calendars depicting scantily clad or nude women and interesting semi-dirty signs. Look demurely at the floor as your husband and the other men studiously avoid staring at the pictures of the unclothed ladies. In the chute office, the potload of calves is recorded; after that the trucks are backed up to an outside chute and the calves unloaded into holding pens.

Following that requirement, trucks are parked in a parking area and the men form into a loose cluster and mosey among the pens, eyeballing and commenting on the critters within. You trail along. From now on you are totally invisible. Yours is not to reason why. Yours is merely to trail along behind. At the sale ring, all climb to the top of the bleacher seats spread around three sides of the auction ring. Keep following. From past experience you know that the men are going to loiter around the yards the entire day. Be patient. Escape is possible.

As the excited, throaty-voiced auctioneer rattles off unintelligible sounds, the men nod wisely. As for you, an occasional clicking of the tongue suffices. Add to that a small shake of the head and you may earn a reputation as a shrewd judge of livestock.

In the ring, the animals keep coming and going through an entrance and exit door located on either side of the auctioneer's high booth. A man

*After you picked it out, clued in the salesperson, and guided your man to the shop door.

with a ten-foot pole keeps the calves circling so all the buyers (who sit down front) can get a good look. Swirling sawdust and flying excrement thicken the air. After a long time of wise nodding and tongue clicking comments on each animal sold, even the men develop a need for change—or they get hungry.

Every stockyard has an eatery always named "Stockman's Café" or, sometimes, "Stockman's Bar and Café." Rows and rows of vinyl plastic, flyspecked booths extend along the walls. The men fit themselves into a booth with you squashed in the corner A waitress, wearing a too-tight pastel uniform and giant dangly earrings, brings a pot of coffee and a stack of cups without being asked. She passes around the cups and pours the first round. After that, the men in the booth handle the job. IMPORTANT NOTE: Pouring the coffee in the Stockman's Café is NOT up to you. Don't even make the attempt! As friends and passersby stop at the booth to visit, your husband grandly waves the coffee pot about and invites them to join the party. HE pours the newcomer's coffee as well as refilling everyone's cup. Lastly, he notices you and bends the spout your way.

At this point, cover the cup with your hand and whisper that you need to exit for personal reasons. Excuse yourself with a lovely smile to the assembled stockmen, tell your man you'll see him later, smile once more, and skedaddle out of there* (see footnote). Head for uptown.

On the way home, you and your spouse exchange stories. While he relates the mysteries of the market, you describe your exciting afternoon. Always show him one of the new shirts you thoughtfully purchased for him. As you draw nearer home, begin a graphic discussion of food that includes juicy steak, baked potato, and strawberry shortcake. Time the gourmet discourse to peak just as you crest the rise leading to the Country Inn Supper Club. (This is a can't-fail recipe to insure a nice ending to the day over cocktails and a lovely dinner not prepared by you.)

*Naturally, as a thinking Country Woman, you have a duplicate set of keys to the truck. Visit the art galleries, buy school clothes for the kids, purchase an outfit for yourself, and have an elegant lunch in a tearoom. (Sometime during the afternoon, buy your husband a new shirt.) Rest assured your husband remains happily glued to the stockyards while you are cramming a year's shopping into three or four hours.

SECTION TWENTY-TWO
Getting in the Wood for Winter

Wherein one learns how to operate the chain saw, throw wood, stack wood, dump wood, and rick wood into neat piles. One also learns how to get the wood hauled from woodpile to stoves.

It's been your privilege to notice that Mrs. Super Rancher lives in a super elegant ranch house where everything matches and is color coordinated. Her house is also uniformly heated by invisible electric or gas genies. One or more artistically designed rock or brick fireplaces with giant mantelpieces above hearths with glass doors draw the eye. The hearths are so big you could roast an elephant on a spit and have room left over. Objets d'art such as original bronze sculptures adorn the mantel and an original oil painting the size of a barn door graces the wall above.

Your own establishment has slightly fewer posh appointments, but does have wood-burning units in more than one room. The major unit provides central heating for the entire house. It's a funny-looking, brown, enameled, metal, square-ish thing about four and a half feet high that takes up an ungainly amount of space in the living room. Rising like a black dinosaur neck out of its backside is a stovepipe, which makes a right-angle turn before disappearing into the chimney. The background behind the stove is not graced by decorative rocks or stone. Rather, painted sheetrock, smudged, merely stands there making a wall. The only objet d'art is a tin can one-half full of water sitting on top, attempting to pass as a humidifier for the room. Draped artistically atop the can may be a pair or two of cotton work gloves drip drying.

The other wood-burning unit in your house is found in the kitchen. It's called a cookstove. It is a dual-purpose outfit that provides heat for the kitchen and the back part of the house, not to mention that during the winter you cook on it to save electricity.

Both the living room and the kitchen stoves eat wood. Lots of it. The wood must be sought out, cut, gathered, hauled, stacked, and eventually carted into the house. Most often this job is one accomplished by your man. Frequently, you get to help with the wood gathering. Sometimes, as need requires, you may go off into the timber on your own. Hitch the small trailer to the small tractor. In the trailer, place a can of gas and a can of chain oil plus your own personal lightweight chain saw your man bought for you last Christmas* (see footnote).

Although the subject of getting the wood in for winter has often been mentioned during the summer and early fall months, serious attention is not paid to the project until the first wet, cold snowfall of the season. Dress and attitude are important in wood gathering. Wear a warm jacket over coveralls, and stout leather gloves. Hat style is optional, but don't forget earmuffs. These last are a must. Earmuffs warm your ears and prevent incipient deafness from the roar of the chain saw. (If you are accompanying your man on a wood-gathering trip, maintain a cheerful air of camaraderie. It's probably the first outing you've had together since haying last summer.)

When all is prepared in readiness for the sojourn, whistle for the dogs, mount your tractor, and sally forth. Arriving at the timber site, dismount, remove your saw from the trailer, and rev it up. To rev up a saw, grasp the pull cord and snap smartly, as with a lawn mower. Several snaps later, as with the lawn mower, go into a helpless feminine flap and get your man to start the darned thing. If he is not with you on the outing, shriek and curse and try again. In awhile it will start or you will have a stroke because your blood pressure has burst a vessel.

Wood sawing is the practice of cutting up big trees and branches into small lengths suitable for the cookstove and the heater stove. Neither great skill, impressive wisdom, nor absolute emotional stability are required to master the art. You will have no trouble. First, locate a nice, not too fat deadfall tree. With your saw growling away, slice off the small, prickly limbs and branches. Then amputate the somewhat thicker branches from the main trunk. After which, commence to make big sticks into little ones. Make the tree trunk into logs. It's best to have a pair of wrap-around goggles of some sort to protect your eyes from flying sawdust. (Usually, you don't have or can't find them. However, a piece of clear plastic wrapped across your own regular glasses and tied behind your head works admirably.)

After a time of sawing, your chain saw runs out of gas. Fill from the gas can you've brought along, check the oil, and fill that, too. Take time out to blow your nose because all that bending over makes the juices flow. If you've been particularly thoughtful and planned ahead, you've brought along a thermos of coffee. Pour yourself some and relax against a tree trunk to admire the sky and scenery. If you're very lucky, you will see several

*Do not scoff. A chain saw is a boon and a blessing to all Country Women. It's a little like the difference between a whalebone needle threaded with catgut and an electric sewing machine.

varieties of birds. Among the treetops, bald eagles swoop and sometimes the flash of a bluebird startles your reverie. After awhile, bestir yourself to charge into the task once more. Use up one more tank of gas in the saw. After two go-rounds you've got a passel of wood pieces laying all over. Commence picking them up and throwing them into the trailer. You can use these moments to exercise and stretch your whole body by alternating arched, over-hand shots with underhand, fast zingers. The extra bends you have to use to pick up the pieces that missed the trailer altogether will remove inches from the waistline.

Finally, refill the saw with gas and oil. Then place it carefully in its case and perch it, along with the gas and oil cans, on top of the trailer load of wood. Climb aboard the tractor and head for home. It's been a zesty afternoon. There is still, however, the problem of unloading, stacking, and toting wood to the house. As you near home, stop singing. Park the trailer-full of wood by the woodpile. Put away your saw and equipment. Go into the house and start supper or have a nice, hot bath.

The second you hear the school bus at the end of the lane, go to the door wearing an excruciatingly warm smile. Indicate the trailer-load of wood and mention to Number One son that a reward of a favorite dessert awaits. Instruct Number Two son to follow in his brother's footsteps. In no time at all, the boys have the wood unloaded and stacked. It's been a matter of but a few moments more for them to tote several armloads into the house and drop in the wood boxes. When your man arrives home from the South 40, invite him to join the three of you for some milk and fattening chocolate cake and ice cream. Remember to grimace slightly and hold your back as if in pain as you get up to pour him coffee. Mention how well the boys unloaded and stacked the wood YOU sawed ALL afternoon. Receive his commendations with a modest smile and offer him more cake. Maybe in the next week or two, you can manage to get away for another afternoon of bird watching and wood sawing.

SECTION TWENTY-THREE

Cleaning the Pig Sheds (Also Done in Spring)

Wherein one learns the important step-by-step method of efficacious shoveling of pig manure.

When the leaves on the cottonwoods begin to think of turning from summer green to gold and tan, all twelve sow ladies begin to dream of baby piglets. The mamma sows waddle as their bellies grow plump.

As in spring, pig mammas also farrow in the fall. Also, as in spring, the farrowing pens must be cleaned and made ready for the ladies' confinements. Although the task is not complex, it has a certain delicacy not suited to all women. Naturally, you enjoy it. Unlike housecleaning, diaper washing, or oven scouring, preparing the pig parlor has a creative purpose. Besides, the pigs aren't critical.

On an auspicious day, attire yourself in bib overalls, a long-sleeved shirt, and your polka-dot baseball cap. Stuff your hair up under the cap or else tie a scarf, turban style, over your locks. It is also useful to wear a bandana around the neck. Pen cleaning is guaranteed to diminish the luster of a Dior outfit. Wear cotton or leather work gloves and take along a thermos of coffee. Important: Load your pockets with a good supply of nose tissue. If you aren't allergic to dust before you start, you soon will be.

Approach pig-pen cleaning with a philosophical whistle-while-you-work attitude . . . at first. After awhile the dust chokes off that sport. Besides, your teeth become coated with grime, making it difficult to slide your lips into a pucker. As the air thickens, raise your neck bandana across your mouth and nose, bandito style.

Equipment for proper pig-pen cleaning consists of a basic, all-purpose pitchfork, a garden rake, a grain scoop, a manure fork, a stiff broom, and a wheelbarrow. Add to these a garden hoe blade that has no handle. It can be noted here that posh pork farms have cement-block farrowing houses with cement floors, air conditioning in summer, and circulating heat in winter. Neat cubicles house each mamma. No untidiness exists anywhere. Somehow posh porkers don't seem to chew up guard rails, bash in feed pans, or defecate on their environment.

Your own setup can be described as folksy, but loving toward fine swine. The building was once a big horse barn in the days before motorized vehicles. The interior is divided by wooden panels into farrowing pens. All twelve pregnant Spot sows are a happy-go-lucky group, but destructive. Happily they'll leap onto a panel and smash it to bits or chomp holes in the walls and floors as they make their nests. (Sow pigs are not to be trifled with when in pre-parturition, nest-building mode.) Each day brings a fresh, broken panel to fix or a hole to board up. The final effect is a mish-mash of uneven boards, jagged-edged patches on patches, and temporary adjustments that remain there permanently. Definitely, your farrowing barn does not qualify for "Swine House Beautiful" or "Martha Stewart Living." But the pigs don't care.

Farrowing pens are constructed like big-city efficiency apartments. The maternity room is at the back, the parlor area, kitchen, and bathroom at the front. At one side of the maternity area, like a gutter space next to a bowling alley, is a narrow fenced-off space called a "creep." The creep allows the babies to cuddle up under a low-hanging heat lamp, but prevents mamma sow from stealing all the piglets' water or baby-pig pablum. A two-by-four guard rail eight inches off the floor and eight inches away from the wall lines the remaining sides of the pen. The guard rail prevents mother from inadvertently squashing one of her children.

When not in use as a farrowing hospital, the rooms are let to just any old pig who wishes lodging for the night. Over the several months since the last farrowing, there has been no maid service. Old, piled-up straw and dirt and other natural refinements lie in thick layers and must be removed.

Therefore, after you get yourself lined out with appropriate tools and equipment, do not dilly-dally. Start with the largest forked implement and scoop up all the loose straw. Develop a rhythm. Singing an old plantation slave song sometimes helps. Scoop up, lift, and toss the treasure into the wheelbarrow. When the barrow becomes full, wheel it to the center aisle that runs between the pens and dump. Repeat as many times as needed.

Later, your husband gets to scoop it all up again and sling it into the big spreader. You can help, but try not to. It's better to exhaust yourself pen cleaning. Or develop back symptoms. Or have a baby yourself. It at all possible, avoid loading the manure spreader. Heaving forkfuls of heavy stuff up above your head, then over and into a spreader is a gruesomely painful chore. And expensive—as often it requires visits to the chiropractor* (see footnote).

*The reason the spreader isn't parked in the central aisle between the pens is that the aisle is too narrow or the spreader is too wide, depending on your viewpoint.

Starting at the back of the pen, scrape down to the bare dirt, then sweep. Assume a kneeling position and use your hoe blade to get at the corners of the creep and pen. The entire pen-cleaning operation, after some practice, should require about an hour per pen. However, time may vary according to how thick the crud lies and how much help you have. Usually several dogs and all twelve sows join you in the pen, busily searching for lost bits of grain.

Beware that a careless pig doesn't step on your hand as you kneel in a corner. Be prepared to scratch backs and rub sow bellies. But try to prevent an affectionate six-hundred-pound sow in search of a belly rub from collapsing full length right where you are sweeping. Eventually the girls become bored with what you are offering for entertainment and depart for a nap. Succeeding pen-cleaning goes a little faster.

Finally, all the straw, dirt, and guck is removed from the pen. Next, go fetch the bucket of barn lime. A scoop at a time, fling the lime about the cleaned pen. Use a wide-armed sweeping motion as though you're pitching a horseshoe or taking a bow. Make certain the lime clings to the upper sides of the pens and take particular care to sprinkle thickly in all corners and along edges.

Last, and very important, close the pen gate and secure it with baling wire at both top and bottom corners. DO NOT depend on new-fangled hooks, bolts, or snap locks. Most sows have a special ability to pick locks and break fasteners. However, there is no recorded instance that they have learned how to untwist baling wire.

Oinked the boar, "This I know will concern ya,
But since I've develop a hernia,
I'm so dad-blamed sore
I can no longer bore,
So I'm forced at this moment to spurn ya."

The sow toppled over the cream
And lay in the widening stream,
Said she, "It's my duty
To care for my beauty,
To look like Miss Piggy's my dream."

A sow sidled up to the wagon
Where the farmer spilled booze from his flagon
She lapped up the liquid
Like any good pig would,
And both of them got quite a jag on.

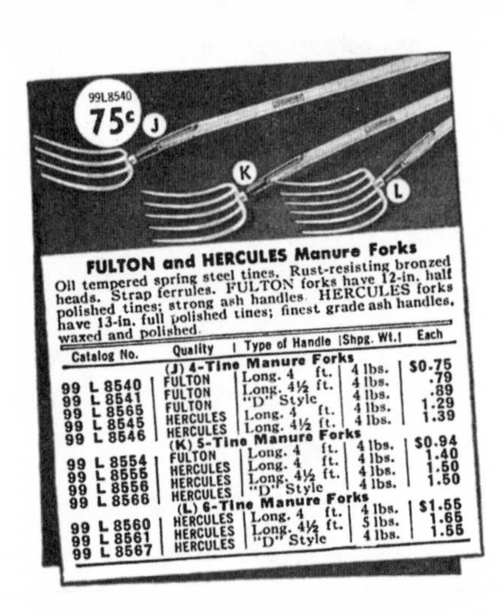

FULTON and HERCULES Manure Forks

Oil tempered spring steel tines. Rust-resisting bronzed heads. Strap ferrules. FULTON forks have 12-in. half polished tines; strong ash handles. HERCULES forks have 13-in. full polished tines; finest grade ash handles, waxed and polished.

Catalog No.	Quality	Type of Handle	Shpg. Wt.	Each
	(J) 4-Tine Manure Forks			
99 L 8540	FULTON	Long. 4 ft.	4 lbs.	$0.75
99 L 8541	FULTON	Long. 4½ ft.	4 lbs.	.79
99 L 8565	FULTON	"D" Style	4 lbs.	.89
99 L 8545	HERCULES	Long. 4 ft.	4 lbs.	1.29
99 L 8546	HERCULES	Long. 4½ ft.	4 lbs.	1.39
	(K) 5-Tine Manure Forks			
99 L 8554	FULTON	Long. 4 ft.	4 lbs.	$0.94
99 L 8555	HERCULES	Long. 4 ft.	4 lbs.	1.40
99 L 8556	HERCULES	Long. 4½ ft.	4 lbs.	1.50
99 L 8566	HERCULES	"D" Style	4 lbs.	1.50
	(L) 6-Tine Manure Forks			
99 L 8560	HERCULES	Long. 4 ft.	4 lbs.	$1.55
99 L 8561	HERCULES	Long. 4½ ft.	5 lbs.	1.65
99 L 8567	HERCULES	"D" Style	4 lbs.	1.55

SECTION TWENTY-FOUR

The Art of Opening Gates

Wherein one learns how to open and close normal gates as well as husband gates.

Sometimes women wonder about their rightful place in the scheme of things. As a Country Woman, you do not have that problem. Your place is at your husband's side—the right side—in the pickup, with one hand clutching the door handle as you wait for a gate to appear. Poised and alert, your eyes constantly scan the road ahead, muscles tensed to spring into action. You are secure in the knowledge that you are NEEDED. As an experienced gate opener you are responsible for tremendous savings of gasoline, shoe leather, and extra man power. The energy conserved because of your special ability would subsidize an entire underprivileged nation.

After awhile you achieve an expertise in gate opening. In consultation with other Country Women* (see footnote) basic techniques have evolved for coping with any style of gate.

Dress code for gate opening is any costume of the moment. However, gloves are a skin saver. Keep an emergency pair in the jockey box of all farm and ranch vehicles.

Gate Type A: Consists of basic metal structures—two styles. Style One: Those slick, neat, aluminum or light steel things. Often painted green, these gates are a marvel of easy opening. To wrest wide such a portal is a cinch. As your man slows to a halt at a gate, assume a casual competent country-person air, slip lightly out of the pickup, and approach the target. Take time to admire the birds and flowers. At the gate, daintily flick upward a little U-shaped gismo that prevents the latch bar from emerging from its slot. Reach for the convenient balanced handle and gently pull. The latch bar slides back and the gate begins to move silently on velvet hinges. Smile and wave at your man as he drives through. Close the gate with the same satisfying élan.

Gate Type A, Style Two: May require considerably more emotional control and muscular ability. Usually, it's a metal unit adapted from a former life such as an old iron bedstead. Frequently the thing is a patched-together assortment of pipe and chain-link units trying to pass as a genuine gate. Always, sharp pieces of jagged metal, wire ends, and weird, rusty formations reach out to slash and tear various bits off you. It is now you will need those gloves.

*Consultation: i.e., gab-session

Undo the gate fastener, which is always a rusty chain or a wound-around piece of baling wire. The gate behaves reasonably well as you grasp, lift, heave, and begin dragging. Suddenly it will spear itself into the ground, stabbing you on the chin as it does so. Stop and spit on your gloved hands* (see footnote). Plant your feet on either side of the gate framework, grasp with both hands low down on the metal edge near you, and then heave upward and drag sideways all at once. As you move, shuffle your feet rapidly to keep pace with the gate. You look a little like a pitiful Egyptian peasant toting a rock up a pyramid.

As your man drives through the aperture, don't look at him. Study the scenery, the sides of the road, or your bleeding stigmata. If you don't catch his eye, and take your time closing the gate, perhaps he'll refrain from that comment about the width and swing of your backside.

Gate Type B: Those standard barbed-wire and peeled-pole outfits. A barbed-wire gate consists of three or four strands of barbed wire stretched tautly between three upright, wooden, vertical peeled poles. The first and third verticals are known as "sticks," while the middle one is called a "stay."

Two general styles pertain here as well. Style One: This stands glistening in the sunlight. The new peeled-pole sticks glow softly. The tight-stretched barbed wire practically hums a tune of pride in its own functional beauty. A wire gate is fastened by a circle of smooth wire attached to the solid set-in gatepost, then looped around the top and bottom of the first gate-stick. The bottom wire loop serves as an anchor and stabilizer. The middle stave supports the three strands of wires as they pass on to the third gate stick, which is anchored with wire to the other solid gatepost. Your job is merely to lift off the wire loop from the top of the first gate stick, pull the stick out of the bottom loop, and drag the gate open.

Wire gates have driven many a fledgling greenhorn rancher or farmer back to the safety of the big cities. No self-respecting gate can be opened merely by lifting the loop off the top of the stick. In fact, no gate your husband has built can be opened at all except by a crowbar or your husband** (see footnote). You, of course, are expected to pinch-hit for a crowbar. The secret is in the use of the shoulder. It's best to be wearing a

*It is not necessary actually to spit, especially if you're not good at it. Just the gesture will do. It's because you have observed your husband and the hired hand doing it before tackling heavy jobs. Remember, you don't have to understand why in order to have faith.

**There are three reasons for tight-beyond-reason gates: one is to keep you alert and strong; another is to keep rambunctious cattle from leaning on loose gates and mashing them flat; the third is to discourage trespassers.

jacket, preferably old and already snagged. Snag damage reaches 87 percent among gate-opening Country Women* (see footnote).

Approach the gate firmly. To cringe is to be defeated. Face the solid gatepost. Slip one arm under the top wire looped around the first gate stick. Reach across and clasp the gatepost in a bear hug. You will now be hugging and squeezing the gate stick as if you've found a long-lost lover. The gate will ease forward enough to let you slip the wire off the top of the stick. WARNING: Sometimes a recoil from the release of tension can occur. Unsightly bruises to the face and body can result from nestling close to whip-lashing gate sticks.

Now that you have the thing open, the easy part is over. Next, grasp the loop-freed gate stick firmly, put your foot on the bottom loop, and pull up, thereby unseating the stick from the wire loop at the bottom. Drag the entire gate either forward or backward. Of course, when you undo the wire loop from the stick, the three sticks and the wire, which have been held tautly apart, immediately collapse in a battlefield mixture. So try to keep the three gate sticks reasonably tight and aligned as you swing open. (Think of being the last in a crack-the-whip line at a skating rink.) Wave husband through the Apian Way. Again, avoid looking directly at him. Stare at the tires. Mention one is low. Then while he steps out to look, you must first stuff stick into bottom loop, then slip the top loop back onto the gate stick. For some reason, two more inches of space now exist between loop and stick and you have to close the gap.

Don't cry out and don't admit defeat. Re-insert bottom of stick more deeply into bottom wire loop, assume the same hugging stance. Squeeze mightily. Scream once, softly. You have caught a hunk of your shoulder between the wire and stick** (see footnote). Squeeze again. Take the fence stick out of the bottom loop and stretch out the gate to try to yank out the wrinkles. Insert stick bottom in bottom wire loop once more. Assume correct gate stance. Squeeze. Close your eyes. Grunt. Pull. Grit teeth. At that moment, over your shoulder, a voice will rumble, "Need some help?" Instantly drop the entire gate. Turn on your heel. Enter your side of the pickup and quickly stop the bleeding and bind up the places where your arm has been skewered*** (see footnote).

*The other 13 percent flat refuse to open gates or are 8 and 3/4 months pregnant or crippled.
**Most Country Women wear blouses with some sleeve, even in summer. This is to cover the ugly purple marks left from vicious gates.
***The thinking Country Woman keeps a box of Band-Aids in all glove compartments.

If your husband remarks that the gate is a "little tight for you," immediately assure him, "Oh, I could have closed it in just a minute more," and change the subject.

Some wire gates are made even more super tight by the addition of a "cheater," which is merely a stout stick, often an old hardwood wagon spoke or a hame from an old horse-harness. A cheater, however, makes it easier to open, provided it doesn't break. The cheater is attached to the gate post by a short length of chain, cable, or extra-heavy wire. The chain attaches at the gatepost and then to the cheater about one third along its length. This engineering marvel utilizes leverage to aid in gate opening. All you have to do is poke the short end (the one-third portion) of the cheater around the gate stick. The longer end is then pushed hard causing a winching action that squeezes gatepost and gate stick together, allowing you to slip off the loop from the top of the gate stick. The only drawback to a gate cheater is that whereas it operates beautifully upon gate opening, somehow on the return trip, the gap-distance is twice as long. You can't get the end of the cheater stick anywhere near enough to lever around the gate stick. There you stand trying to force two polarized sticks towards each other. Only now you have one more stick thing in your hands to sort out.

The thinking Country Woman will have her own handy portable cheater stick stashed somewhere out of sight. Behind the seat of the pickup is good. Paint your personal stick pink. Your husband isn't so apt to throw it away if it's pink. If he questions you, mention something vague about a craft project you're working on. Wind a good amount of all-purpose baling wire around your stick for added insurance. Thus you can be prepared for any gate. It is recommended that personal cheater sticks be employed only when you're alone. If your man is along, go into feminine flap and he will come to your rescue. It is on those lonesome forays to the South 40—when you can't get the gate shut and will be killed if you dare leave it open—that your pink cheater stick becomes a life saver.

Gate Type B, Style Two: A great deal similar to the first style in basic construction. But instead of being super tight and impossible to open, these have broken sticks and/or broken middle staves held in suspension by the rusty wire clinging to them. Although there is no special muscle prowess needed to open these antiques, closing them can become a new dance step. Loose wire reaches barbed tentacles to entrap and ensnare. It's best to keep the whole mess pulled full length to prevent the equally shaky fence from collapsing.

Gate Type C: Includes those big, heavy, wooden corral gates in a variety of sizes. They are very functional and very heavy, which is perhaps why they always sag. The fastener is a wooden bar which shoves into slots in the gate post. As the gate sags, the slot stays where it is while the bar moves downward with the tired gate. Corral closures such as these cause hysteria in a Country Woman and evoke berserk rages from your Country Man because you don't get them shut or open fast enough. Naturally, it's your fault if a stampeding critter gets away.

Before really severe sag-it-is sets in, you can usually manage by lifting firmly on the gate while pulling back on the bar. As time marches on, the gate drops more. Eventually, there are about two zillion pounds of weight, which must now be heaved upward each time you attempt an opening.

About then, the time arrives when it's your job to swing open the gate for an advancing group of heifers being shagged into the corral by your horseback husband. As the cattle close in, move smoothly to your appointed task. Except you can't budge the gate. Closer and closer come the critters. Frantically, you lift, heave, and haul, but the only thing that moves is that slipped disc in your already damaged back. Finally, the heifers begin to scatter away from the gate* (see footnote). Your man rides forward, swings off his steed—just like in the movies, makes several declarative statements using terse four-letter words, jerks open the gate, remounts, and rides off again to re-roundup the heifers.

At this point, turn on your heel, go to the house, march to the basement, and fetch your very own chain saw. Stalk back to the corrals, kneel down, start the saw, rise, and commence cutting a new slot below the old one in the gatepost. Pay no attention to what's happening in the corral, and under no circumstances speak to anybody or open any gates until you are satisfied you can.

*Cattle tend to react with terror at the sight of a medium-size plumpish woman moaning at the gate.

THOSE DAMNED WIRE GATES

The sun was high, the weather fair,
As I roamed the hills on my buckskin mare;
The ride was long, I was running late
When I pulled up short at a barbed-wire gate.

Now I know gates and I know they're mean,
But the ranch-house roof could be plainly seen;
If I went around, it was five miles more,
And my seat and my knees and my back were sore.

Once I asked my spouse, "Why are gates so tight?"
He looked at me like I wasn't bright,
"Cuz a bull or a cow can lean their weight
And knock the wires off a loosened gate."

I slipped from my horse and I faced my foes—
I would at least strike an opening blow
I tried the top, then the bottom band,
I tore my sleeve and I hurt my hand.

I wrapped both arms around the post,
I pulled till I ran out of breath—almost—
I kicked its wires to show my grudge,
I cussed it soundly—it wouldn't budge!

I checked my pack for a tool for prying,
A metal bar would be well worth trying;
But the only bar was a candy one
And even with peanuts it wouldn't have done.

The Time was approaching six o'clock,
When all of a sudden I spied a rock.

With gusto and grit I began anew
And hammered the lower wire in two.

With great relief I mounted my mare
And left that gate just lying there;
Glad to leave the scene of the crime,
We trotted home in record time.

This morning early, my other half
Said, "I was out checking a newborn calf,"
And then he grumbled, "Guess what I found—
The south-field gate was on the ground!"

I answered then and my voice was cheery
"You know, when I rode out there, Dearie,
Your black bull was quite a sight
Leaning against it with all his might!"

Now I stretched the truth, which I deplore,
But if everything's fair in love and war,
I'll be forgiven by all the fates,
Cuz I'm at war with those damned wire gates!

SECTION TWENTY-FIVE

Hunting Season

Wherein one learns how to cope with those odd creatures—hunters. One learns what to tell them and what not to tell them. One also learns the correct method for helping husband butcher game meat.

In the fall, along with Nature's glory, come the HUNTERS. When the leaves turn gloriously golden and the deer from brown to gray, you know it's fall. You also know HUNTERS are on the way.

A primitive something strikes deep into the hearts of male persons—especially male city persons. They swarm into the countryside driving amazing motorized transports. Many and marvelous are the variations and combinations of hunter vehicles. Pickups with campers pull trailers; mobile homes pull jeeps; jeeps pull snowmobiles; converted school buses with flowered curtains at the windows pull horse trailers. Some remarkable outfits manage to combine all in one wonderful conglomeration of motorized, expensive extravaganza, such as a block-long motor home, pulling a flatbed trailer, which carries a jeep, a snowmobile, a canoe, and a kayak.

The interiors of the remarkable conveyances are equipped to the eyeteeth with all manner of hunter appurtenances, such as electric socks, orange hunting vests, booze, jackets with special ammunition pockets, booze, special ropes for tying on of game, booze, and red and orange plaid hunting caps. And booze. On some mornings during hunting season, it's possible to hear a hollow thwacking that sounds like someone hitting a rotten log. Actually, it's the echo of manly chests being thumped as armies of HUNTERS prepare to go after fierce, wild meat.

City hunters spend a good deal of their time wandering up and down country back roads, searching for a PLACE TO HUNT. Most of the road wanderers stop and ask permission to hike on the farmer's or rancher's land. Usually they rap on the door about five or six a.m. Sunday morning. (They don't hesitate to bang on your door at that hour because they've heard positively all farmers and ranchers rise before dawn.) At whatever hour they appear on your stoop, you and your husband have a set of prepared remarks.

However, try to be kind the morning you are trudging across the barnyard and look up to see a remarkable apparition. A club-cab pickup with camper shell pulling a small, elegant house trailer behind itself is tiptoeing up your lane. Hitched to the house trailer is a miniature two-wheeled trailer upon which rides matching snowmobiles. Atop the camper, perched like a cap on a schoolboy, clings an upside-down canoe. Clamped to the rear of the pickup sits a pair of low-slung dirt-bike bicycles, the spokes winking in the early rays of the sun. The assemblage stops when it sees you.

You stop also. You expect at least Pancho Villa, but what emerges is a mild-looking, slightly paunchy individual who appears distinctly nervous. Since you carry nothing more intimidating than a grain bucket in one hand and three eggs in the other, and are followed by two peace-loving dogs, a cat, a buckskin mare, and two medium-size sows, you are sure it can't be yourself making the man edgy.

It turns out he wants to know if the timber brakes along the creek bottom is a "wilderness area." It so happens that's the very pasture into which your husband has placed the weaned calves.

Instruct the man to come back later when your husband will be home. Mention he's not expected in until dark because he's tracking down three rabid coyotes that have been ripping up the countryside. Or tell him the game is all in the high mountain country and give him the name and phone of a local guide. Or whip out a map of the local forestry reserve land and point out the "real wilderness" where crowds of deer abound. Lastly, offer to sell him a steer calf and promise to throw in a set of antlers from your son's antler collection.

Local farmers and ranchers who live in the country all year round do not take as many pains with hunting. When your husband goes after winter meat, he dons his cap with the earflaps (flaps in the up position), checks out his .30-.30, drives to the upper section, quietly stalks downwind from a herd of grazing deer or antelope, and picks off a buck. Usually he's back inside of two hours.

As a good Country Wife, you needn't feel left out on any possible glory. If you like to hunt, go for it. If you prefer to remain behind ready with applause and coffee, you're a member of a large club. Either way, as the mighty husband hunter returns with his kill, immediately leap into the role of helpmate.

Your spouse hangs the dead creature from a rafter in the old smoke house. After awhile, he skins it (the animal, not the smoke house). While he's skinning and removing the head and other unneeded parts, you are busy in the kitchen spreading the table (after inserting extra leaves) with a piece of plyboard and arranging the meat saw, butcher knives, and freezer wrap. All you lack are strobe lights for the coming operation.

Dissecting the animal is a necessary job especially if you want it to fit into the freezer. A competent meat cutter can whack up a critter into steaks, roasts, and chops in a short time. You are not that good. In fact, the whole process is messy, bloody, and very tiring. Therefore, it's best if you do not become proficient—ever—in cutting up meat. But certain steps may be taken to help maintain your image as a true Country Woman. Wear a long canvas or rubber butcher apron or an old shirt of your spouse's over your own clothes. If you don't like the gucky parts of the meat-cutting, wear rubber gloves. Have two large, cardboard boxes lined with giant garbage sacks ready for meat scraps. Place pieces of cheese cloth and string over the back of a nearby chair. As your husband cuts away the suet, wrap and tie it in pieces of cheesecloth about baseball size. Later, hang them in the apple trees for the birds.

During your preliminary preparations, glance out the window from time to time. As soon as you observe your man heading for the house with a quarter of a deer over each shoulder, turn on the television loud enough to hear it in the kitchen. Next, make yourself and your man a tall spiritual beverage.

As the cutting chore progresses, keep freshening the beverages (or yours at least). Your man will be working so hard sawing through bones and chopping away the hard parts, he will sweat out any alcoholic effects, especially if you've watered his beverage. Since you are only wrapping the portions and marking the contents with a felt marker, you may develop a more severe reaction (since you have *not* been watering your beverage). By the end of a long meat-cutting session, you should be so comfortable, all the blood and gore won't bother you a bit.

If you are not a squeamish meat cutter, you are a True Blue Farm or Ranch Woman. You know neighboring Country Women who have no qualms and happily butcher all the meat animals—steers, sheep, pigs, or game. But remember, not everyone can qualify for Best All-Around Country Woman Award. Somebody has to be among the runner-up group. Anyway, it's a lot bigger club.

TROPHY ELK

The Mighty Hunter came out from the East
Intending to hunt for a wily, wild beast.

An antelope, a deer, a moose, or a goose,
Or maybe a bear or an elk on the loose.

He had purchased his tags a long time back,
He had so much gear, he could barely keep track.

Against his wife's wishes, he bought for the trip,
A brand new Jeep—and she nearly flipped.

But he had to do it, he was on a quest,
To hunt the wild beasts, he must have the best.

He couldn't be faulted for the way he was dressed,
In a loud checkered shirt and a bright orange vest.

And canvas pants of camouflage brown,
And sturdy brogans and a jacket of down.

By gosh, he was ready for the hunt of his life,
He was armed with a gun and a long sharp knife.

Over hills, over dales, across streams, and through brush,
Onward and onward and onward, he mushed.

Though he 'scoped the plains and the shadowy rims,
Not a critter did he see, and he grew rather grim.

His spirits they drooped and his hunter's lust,
Began to diminish while his Jeep acquired dust.

Then suddenly, there! Way off to the right,
He spotted an elk through the telescope sight.

It was big, it was brown, it had a high head,
As he drew a tight bead, he suffered some dread,

Fearful he might miss this huge trophy elk,
Why, think of the stories he'd have to tell

To his buddies back east! And his wife would be glad;
When he brought back the meat, she'd stop being mad!
He squeezed the trigger and the bullet flew straight,
The animal fell over; it had met its dire fate.

And the Mighty Hunter got tears in his eyes,
Till through his binoculars what did he spy?

A cowboy approaching that fresh-killed elk;
Mighty Hunter trembled from the fury he felt.

He leaped in his Jeep and sped to the site,
And piled out of the rig prepared for a fight.

His anger boiling over, he aimed his rifle;
But the cowboy seemed to be weeping a trifle.

But Mighty Hunter showed no mercy,
"Get away from my elk!" He started in cursing.

The cowboy backed off, saying, "Mister, you're right,
This sure is your elk and I don't want no fight."

"But I have one request, and I swear it ain't idle.
Please let me remove that elk's saddle and bridle."

PART FOUR

Winter and the Layered Look

Winter: That time of year when one huddles a lot. Mornings are a misery if your sheepskin slippers have disappeared. And if the pipes have frozen in the night, that trip to the outhouse is pure agony.

SECTION TWENTY-SIX

Feeding the Critters

Wherein one learns to drive the tractor or the pickup at snail's pace while your man throws bales to the cattle. One also learns how to throw the bales when husband throws his back out again.

Sometimes fall and Indian summer last all the way to Christmas. More often the blizzards strike early. A crisp, brilliant, clear sky begins to cloud up. The air grays and thickens, and before long, that thick air congeals into gorgeous snow. The thing about snow is that it's everywhere and all over. Not a twig escapes the cold, white blanket. While you may gaze in awe at the incredible beauty God has wrought over rooftops and fence posts, your man is apt to use the Deity's name in vain.

Snow and cold mean the cattle must be fed hay. The very hay you personally helped stack must now be dispensed, a few bales at a time, to the hungry critters.

Your man prefers to feed at daylight (naturally), which means you are required to rise in the middle of the night in order to feed him first. Also, naturally, you are privileged to help with the chores.

Gauge the layers of your outdoor attire by the outdoor temperature. The colder it is, the more sweaters, vests, and warm socks you must don.

CAUTION: Avoid squashing yourself too tightly into a many-layered outfit. The circulation can be squeezed from a limb in no time. The pain, when the blood finally begins circulating again, is excruciating.

When feeding in very cold weather, the following toilette procedure is recommended: Circle that long, red, woolen muffler around your neck, and up over the hood of your old, red snowsuit. (Dudes from the City call a snowsuit a "snowmobile suit.") Wear your husband's old vest under the snowsuit and his old denim jacket over the outside. As mentioned, the trick is to avoid tourniqueting any part of your body. On your feet, wear the usual high-laced boys' boots over two pairs of thick socks. Lastly slip on large, four-buckle overshoes and buckle the tops over the pant-legs. You will look like a displaced, pudgy Eskimo. Try to avoid looking in a mirror or having to use the bathroom while thus outfitted. Lastly, feel your pocket to be sure you have your pocket knife, take up your warm, fleece-lined gloves or mittens, and stagger outdoors. When King-the-Collie rears up to greet you, make sure you are standing near a wall. Being bowled over in the snow on a cold, dark morning is funny only in comic books.

Your man has not been idle. While you have been garbing* (see footnote), he has already loaded the pickup or the tractor's flatbed trailer with bales of hay. Whether you use the pickup or the tractor to feed depends on which pasture you're going to, and how deep the snow lies. As a good Country Woman, your job is to drive the vehicle while your man rides behind to throw off the bales. He will designate where he wants you to drive by waving his arms and pointing—similar to the way sheepdogs are directed to maneuver after sheep. You must keep one eye backward to watch your husband's signals** (see footnote) and one eye forward to avoid falling in the irrigation ditches, sinking into a drift, or hitting a cow. Never go too fast. A lurching crawl, like a gluey-footed centipede, is the idea.

If you are driving the pickup, your man will yell, "Put it in the hole!" This means you are being ordered to utilize the very lowest gear—called compound*** (see footnote). The tractor also has an extremely low gear whose location you can't remember. Unless you get really stuck though, a tractor can putz along as slowly as you wish even in higher gear.

*No one knows why it takes a woman three times as long to put on her outdoor gear than it does a man. Try to get a head start dressing so you won't be left having to walk to the haystack again.
**When you can't see the signals due to the height of the load, you must hang one ear out the window and try to interpret the shouts. Attempts to use ESP do not work.
***It doesn't matter what it is called. The important thing is to be aware that when "in the hole," the outfit will inch along over rocks, logs, and up the side of a tree if pointed to it.

Eventually, your man will drop the last bale and signal you to head for home. It is at this point many Country Women lose face. As you shift from compound, do not jerk. Otherwise, you may lose a good man off the tailgate. If you notice he's missing, be sure to stop and pick him up. He'll be a lot madder if he has to walk all the way home.

There are those cruel mornings when it's entirely up to you to throw the bales because your man has put his back out. But he can still drive, although painfully. Ask him to back the outfit smack up against the haystack, then grab a bale hook and climb to the top of the stack. Many Country Men disdain bale hooks as nuisances when loading a hay wagon. Pay no attention. Bale hooks are a boon and a pleasure to the Country Woman. The hook bites satisfyingly deep into the bale. You won't have to wear out fingers and wrench shoulder sockets trying to hang on to those bales by the strings. With a hook, you can drag a bale forward and then lift and heave. Be sure to unhook before you heave, otherwise the hook stays lodged in the bale and follows it down to the haywagon. Worse yet, you may still be hanging on to the bale hook handles* (see footnote).

When the hay load reaches the proper limit** (see footnote), jump on top and hang on while your man hurtles the vehicle forward to the feed ground. As he puts the outfit into its slow crawl, grab a bale by one of its two strings and give a mighty upward wrench while kneeing the center of the bale at the same time. Theoretically, the bale string will pop off like a rubber band. Then holding the one length of string, reach down and grab the other string, still around the bale. A quick flip and the bale flops apart and scatters over the edge of the wagon bed. Repeat this with additional bales at regular intervals, first on one side of the outfit, then the other. Each time after dropping a bale, you will be holding two strings, which you then loop over a handy upright pole on the pickup. Later, you will burn the strings.

Unfortunate truth: The theoretically correct method of bale throwing only works some of the time. More often, the blame things fight back. Bale strings refuse to pop off no matter how hard you pull. If that happens, slip off your mittens and hold them in your teeth or, if you don't care for the slobber, stuff them in your back pocket. Dredge out your pocket knife and cut those reluctant bale strings.

*At this point, your man's sense of humor asserts itself. He laughs himself silly.

**Try to keep mouth shut when your man comments on the flotsam and jetsam appearance of your load.

Some sort of implement for cutting strings is a must. A tooth from a mower nailed to a stout stick works very well. Hold the tool aloft and bring it sharply down slicing the strings like a guillotine on the head of a wife of Henry the VIII. An old hunting knife is the choice for some. Stick it, point down, in that skinny pocket half-way down the right leg of your snow suit. When needed, you can whip it out and whack the strings apart. (This particular method has a certain Wild West style to it.)

Sometimes a string will break as you reach for it, scattering hay all over the load, but not onto the ground. Or you fall down, lose the whole bale over the edge with strings still tied, and are forced to holler for your man to halt. Then jump down, run back, and break that bale. Tear back to the outfit and haul yourself aboard, which is terribly hard to do in all those clothes, especially as the vehicle always starts forward just before you really arrive.

Finally you finish the job. As your man turns toward home, remain standing on the truck or tractor bed. Clutch firmly anything available. Do not make the mistake of seating yourself on a bare board floor when returning from the feed ground. To do so will invite painful, long-lasting bruises as your husband guns the vehicle over the terrain, each wheel hitting a different frozen pie.

As soon as you reach the house, dash to the telephone and call the bone doctor. Plead mightily for an immediate appointment to have your husband's back overhauled, because your own is rapidly fusing into a permanent L shape.

If you don't create square bales, you and the bank may be the proud owners of equipment that makes those huge round bales. Big machinery sucks up the hay into round pegs the size of a tunnel big enough to drive through if they were hollow. They aren't. To feed these behemoths requires a truck outfitted with a gizmo that has robot arms. Push a button inside the truck cab and the arms will pluck up a round bale and bring it onto the truckbed. Sort of the way a frog laps up a fly only not so quick.

If you, the True Country Woman, are placed in the position of feeding the cattle on some morning or other, no problem. The bales hit the ground and all you do is clip the strings and push and the bale unrolls like a rug—usually. If it doesn't, then you must jump down and pick a fight with the monster.

COUNTRY WOMAN'S PRAYER

Now I lay me down to sleep,
I pray the Lord my soul to keep.
If I should die before I wake,
I pray the Lord my soul to take.

In my youth I learned this prayer,
Rhymed and metered, soothing care.
But what about the daytime, Lord?
When hazards cannot be ignored?

Living on a farm or ranch
Means every day's another chance
That Fate might deal a loaded deck,
And I'll expire in some bad wreck.

Consider, Lord, the many days
That livestock I've turned out to graze
All try to hasten my decline
By accident or sly design.

A calf, oh, Lord, might kick my shins,
Its mama knock me off my pins.
If she don't kill me on the spot,
Dear Lord, she'll cover me with snot.

And Lord, sometimes I have to pill
A scoury calf that's mighty ill.
I sneak up silent as a cat,
My heart a-goin' pitter-pat.

I sit on top and hold his head
He won't budge, he looks half dead.

Big surprise! He leaps straight up!
With me astride, he starts to buck.

Two mighty jumps, and I am flung
Splat on my face in piles of dung.
I check myself, ain't nothin' broke,
Oh, Lord, I cry, is this a joke?

When time to feed the cows again,
Power tools are my best friend.
I load round bales with metal arms,
Cab-controlled, the gadget charms.

I halt the truck to swing the gate,
My luck runs out, I'm slapped by Fate.
The bale truck's brakes have failed to set,
The fence is now a holding net.

I cut the wire, jerk tangles free,
I scrape my arm, I snag my knee.
Rainbow bruises, gashes and cuts,
My town friends, Lord, think I am nuts.

But onward now to feed the cows,
One more prayer if the Lord allows,
For when a bale is dropped to ground,
It should unfurl like a rug in town.

It sometimes sticks and scoots along,
Like a dragging log; Lord, something's wrong!
I put 'er in neutral and jump from the truck
To hand-push the bale till it comes unstuck.

Needles of hay stab my hand,
Oh Lord, I've had about all I can stand.
The cows are impatient, they're in my way,
They don't give a durn 'bout people who pray.

But I've three more bales I still must feed,
So, I ask you, Lord, to hear my need.
It's a modest request, it's a simple goal—
Oh, *please* dear Lord, let my bales unroll!

SECTION TWENTY-SEVEN

Proper Winter Clothing

Wherein one learns how the layered look is achieved for different temperatures.

Unless you farm or ranch in a warm zone, several months of the year must be devoted to the application of outdoor clothing to bodies. It is the Country Woman's obligation to keep track of jackets, boots, gloves, earmuffs, scarves, mittens, stocking caps, long underwear, and fleece-lined winter caps with ear flaps.

In the big, white house owned by Mrs. Super Farmer the ideal storage organizer of winter wraps is a roomy group of closets and cupboards painted sparkling white, which never get dirty or muddy. Inside, all jackets and coats are required to hang neatly on hangers. Overshoes and snow boots line up super neatly in personal grottos of a built-in, polished, redwood shelf unit. A clever drawer section of the redwood unit houses mittens, gloves, and scarves.

Your own system may not be quite as perfect. Two long, paint-chipped boards extend the length of one wall on the enclosed back porch. One board/shelf is above your head. The other is flat against the wall about kid high. Marching across the flat-on-the-wall board are hooks, and when those run out, giant nails. Jackets, coats, slickers, chaps, and irrigating boots hang there like bushy growths of some weird plant form. Beneath the forest of apparel is a long, wooden box that once had a motor shipped in it. Rubber boots, four-buckle overshoes, snow boots, riding boots, one irrigating boot, and two pairs of old, holey tennis shoes spill over the edge. Mittens, gloves, scarves, and stocking caps are supposed to be kept in two of the three drawers of an old commode. (The third drawer won't open and has no bottom anyway.) However, all the items as they are used are tossed in a jumble on top. There's an unwritten law, which prevents putting away any of these things. Eventually, there's nothing left inside the drawers, except for three left-handed mittens. Everything else is piled in a mountainous hodge-podge on the top of the commode (an antique, by the way, that antique hunters would drool over).

Glove and mitten hunts can become a tradition with its own rituals. There are stirrers, pilers, and scoopers. Stirrers slosh everything in a circle till they see something they like. Pilers stack everything like a hero sandwich and then try to find the perfect match. Scoopers merely shovel up a loose heap of gloves and mittens and let them rain down. Eventually they fish out two more or less equal mates.

Basic cold weather clothing begins in the fall with a light jacket over ordinary clothes. Progressively, additions will be made till the really cold weather clothing will include long underwear, warm shirt, sweatshirt, vest, snowsuit or coveralls, and top jacket—lined. At least two pair of wool socks are worn on the feet, inside the high-laced boots. Four-buckle overshoes are last. In zero- to fifty-below cold, you can put a pair or two of your husband's big, olive-drab work socks over the high-laced boots and then add the four buckles.

The last item, although not technically clothing, is truly of utmost importance. Throughout the pockets in the layers of clothing, stuff gobs of tissue. It is gruesome to have a continuously runny nose and be a mile from home. Your husband's suggestion about using a finger alongside a nostril should be ignored. You'll just get it all over you and make your sinuses hurt. Plus, it's disgusting. Yuck.

With each additional layer of clothing, your agility is somewhat decreased. Finally you must be cautious or you will fall and roll away somewhere.

Husbands resist winter clothing. It has to be at least zero before a Country Man will admit his ears are cold. Although he will usually add a vest under, and a light jacket over, his unlined coverall, his neck remains naked to the elements all winter. Whereas you are swathed to the eyebrows and beyond. In the very coldest winter, a ski mask with slits for mouth and eyes is excellent. If the wind is also blowing, add earmuffs.

During the winter months, be increasingly cautious when walking about, especially when feeding in the corrals and pig pens. When it's cold, all those fresh piles (and there are frequent new ones) tend to freeze immediately. Therefore, you must develop an adaptive gait. Lift your overshoed foot straight up, then forward and down—like the moon-walkers do. Should you get careless and walk normally, your forward-moving foot will thonk into a frozen pig pile. The best that happens is a jarring impact followed by an irresistible forward-leaning hippity-hopping momentum till you can catch your balance. Too often though, the impact crashes you to the ground. The grain buckets you are, of course, carrying, upend themselves or slam painfully into you, or both. It's all right and probably therapeutic to curse and scream, but it's best not to linger there, moaning. In spite of the pain, rise immediately, or you will have several swine trampling up your spine.

HATS FOR ALL SEASONS

(This poem was recited by the author on the Tonight Show *with Johnny Carson, along with a show-and-tell of the different chapeaus. It was scary.)*

Those high-crowned hats with mile-wide brims
Are the Hollywood cowboy's glory,
Make ugly men handsome and short ones tall,
But that's not the whole of the story.

On the five-point antler behind my door
Hangs a potload of western hats,
All shapes and colors, all stages of wear,
They cling like a bunch of bats.

Now hats for chores tend to be caps
Emblazoned with feed-store names.
I like 'em a lot when slopping hogs,
Or pursuing similar aims.

At summer's beginning I drive into town
In a straw chapeau unrumpled.
By summer's end, the dust and the sweat
Turns the poor thing soiled and crumpled.

When I ride in the rodeo parades
Decked out in Western style,
My Stetson sits at a rakish tilt
I salute the crowd with a smile.

Those times when weather turns to wet,
And I haven't a slicker nearby,
I grab Old Gray that's way too large;
Odd, I admit, but I'm dry.

As temperature cools and winter nears,
My headgear acquires savoir-faire;
With ear-warming flaps for my chilly ears,
I know I am plumb debonair.

Then blizzards howl, the mercury drops,
And to keep my noggin warm,
I double and triple what I wear on top,
In true-grit cowgirl form.

But spring comes again and the feed-store cap
Rounds out the circle full;
To movie cowhands whose hats never age ...
I say you're full of bull.

SECTION TWENTY-EIGHT

Getting Ready for Thanksgiving and Christmas Food Consumption

Wherein one learns to be grateful for all that stuff one canned and froze. One also learns how to shortcut and come out smelling like a gourmet.

The calves have been sold, the weaner pigs have been sold, the land payment has been made, and the human offsprings are settled into the routine of school. The weather has held nicely. Really foul storms have not yet descended. Number One son has actually done his homework several times without being reminded. Grab all the bits of peace while you can because the Holidays are coming.

The good Country Woman is required to bake, cook, and create vast amounts of foods forever. It's a bad mark on your Country Woman's merit page if you go out to a restaurant for a holiday dinner or have anything catered or patronize a delicatessen. Since the last two aren't available in the country anyway, you have only the first item as an alternate choice to cooking. Even that may be taken from you since restaurants in country towns are apt to close during the major holidays like Christmas and New Year's.

Organization and stamina are keys to success. During this season of the year, try to avoid colds and flu. Even if you catch something, you are not allowed to take time out to be sick. Over the years, a thinking Country Woman will evolve an attack plan to cope with the Holidays. Some women develop a serious lassitude and have to be sent to a rest camp in Arizona. However, you are not financially prepared for that. And since the family won't pay attention to your symptoms anyway, you may as well roll up your sleeves.

Begin by dividing the work and activities into manageable segments. Segment One is food. You will need Thanksgiving Food, Christmas Food, New Year's Food, and Fancy Tidbits Food such as cute cookies and little stuffed pastry things for drop-in company.

First, pick a week around the middle of November. On day one, descend to the basement or tromp out to the root cellar taking along a big basket or box. Fill it with an assortment of canned fruits (especially apples), vegetables, and anything pickled. Back in the kitchen, search out the grinder, the chopping board, and the bread board. Have plenty of flour and sugar on hand.

In the freezer you have turkey, chicken, and roasts you've been saving back. These present no problem besides drudgery. It's on the wide variety of fattening foods you must concentrate. Prepare a basic group of fancy breads, cakes, and cookies. Think in terms of six batches of everything.

Stir up batter for six loaves each of two kinds of easy quick breads, such as carrot or applesauce. Add to each of these raisins and smashed nuts. When the mixture tastes pretty good, pour into loaf pans and stick in oven to bake. Always bake at a slow temperature. When still hot from the oven, pour on a lemon glaze. After the loaves cool, wrap individually in foil and freeze. That's one day's effort, and you are a little smug.

On day two, stir up a basic white cake and a basic spice-applesauce cake recipe. It's best to double these recipes three times. Trying to stir up a six-times batter won't work. You don't have a big enough bowl unless you use the canner tub which you might not be able to lift to pour out the batter. To each of the cake recipes, add a lot of raisins, currants, dried fruits, and some of those smashed nuts. Then add a half cup of liquor—any kind. Bake in loaf pans (fill only half full) at moderate temperature for a long time. Just before you stuff them in the oven, arrange maraschino cherry halves in artful designs over the tops. Or wait till the baking's done. The cherries look smarter uncooked.

When done these "fruitcakes" can be frozen till you are ready for them. Or they may be wrapped in cheese-cloth soaked in booze and placed in airtight containers. Three-pound coffee cans are good. When you think about it, moisten the cheesecloth from time to time with brandy or some other liquor. When the time comes, the fruitcakes can be served with a rum or brandy whipped cream. If you don't have the real rum or brandy, use vodka or gin with rum or brandy flavoring in it. No one can tell the difference. Be modest, but evasive when people exclaim over your superb fruitcakes and beg for the recipes. Mention (reverently) a pioneer ancestor who passed on the recipe with her dying breath while fighting off the hostiles. Baking the cakes along with all that tradition will have used up another day.

On the third day, stir up a big batch of yeast bread dough. Use buttermilk and several eggs in the batter. After the dough has raised and you are ready to work it, do anything except make plain old loaves. Work out a variety of shapes. Spread the surfaces liberally with butter, sprinkle generously with cinnamon, raisins, currants, and some more of those nuts. Leave some plain and some with just celery seed sprinkled over. Then roll, twist, fold, or sandwich the shapes together and automatically you have five or six kinds of dinner rolls. Arrange these in foil pans and freeze. When needed, put them on the heater to raise, then bake when ready. After baking, frost some with interesting colors of icing for "sweet rolls." Making the rolls requires another day. By this time, you may find your enthusiasm waning.

Cookie baking and tidbit fixing could take up the rest of the week. You are a fortunate woman if you love to bake cookies and little candy things. In which case, leave this to last and best. However, if by this time you're sick of the whole business, it's time to cash those cream checks you've been saving. Then call the lovely Norwegian lady who does specialty baking as a hobby. Order six dozen of everything she makes. There is, however, no need to chatter on to anyone about this.

Merely mark off day four on your calendar. That afternoon, sit down and read from two till four p.m.* (see footnote).

On the fifth day, devote yourself to making jellies. You have lots of canned berry juice in the root cellar. All year you've saved and collected small jars of assorted shapes, especially "cute" ones. Baby food jars are excellent if you are going through that stage. (Somebody in the area will be!) Make as many different varieties of jelly as you have kinds of juice. Do one recipe batch from each type juice. (You'll have odd amounts of different flavors left over.) Fill the jars with the jelly and don't worry about more of one kind than another. Eventually you wind up with five or six varieties of teeny jars of jelly. Seal with paraffin and mark the name of each one on those Christmassy Santa or angel stickers and paste it on the appropriate jelly. Lastly spray paint all the jar lids a bright color. And if you're into adding an extra touch, tie on a bow of raffia.

Jelly-making day is a tough day, but when finished, you have all those containers of actual homemade jelly to give away to people who stop by at Christmas** (see footnote). Package them in variety packs and send to city friends who have everything. Take some to the Woman's Club "Bring a Gift" Christmas program.

Combine all the leftover bits of juice mentioned previously into one pot. Using a standard cup of sugar to a cup of juice recipe, make a final

*Do not be tempted away from your reading program by housework or any emergency short of broken bones.

**City people always remark, "You MADE these YOURSELF!? When did you find time?"

big batch of jelly and pour into large jars for the family and label "Fruit Surprise."

On the sixth day of what has shaped up to be a very long week, make candy. Some women are candy-making experts. They have secret recipes handed down by the original candy witch. These ladies go to marvelous lengths to produce creams, mints, bonbons, and unnamed confections of all sorts. However, if candy-making is not your bag, it works best to follow the recipe for fudge on the side of the cocoa box. Use real cream instead of milk and stir in a lot of extra butter and you can rarely miss. Think in terms of four. Make four batches of fudge, two plain and two with nuts. Prepare ahead several platters, plates, and pie pans heavily buttered and layered with rice crispies. Use two batches of fudge over the rice crispies. Spread the last two batches into plates or pans. Sprinkle the tops thickly with peanuts and/or coconut. (This is especially good for those people who like to pick nuts out of candy.) Always buy a jar of maraschino cherry halves at Christmas. On some of the candy pieces perch a half cherry for that professional touch.

If you feel you need another color of candy besides the basic chocolate brown, make a penuche and do all the previously mentioned variations or make up some of your own.

An assortment of candy spread on a paper doily on a pseudo-crystal serving dish looks terribly posh, especially if you add some of those Norwegian tidbits you ordered on day four. WARNING: Do not, under any circumstances, leave unprotected a confectionary arrangement, especially if you're planning it for company. Children, cats, dogs, and husbands become notoriously untrustworthy in the face of such temptation. It's best to hide the dish. Take care no one is watching when you stash the treasure.

On the seventh day—rest.

ROSE-HIP SYRUP AND JELLY

Wild rosebushes grow most everywhere. The red berry that develops on the tips of the branches in the fall is a "rose hip." Pick a bunch of rose hips—carefully. They're stickery. A thin pair of leather gloves can protect your hands.

TO OBTAIN JUICE

Wash and cover rose hips with water in a large kettle. Boil till pale in color and simmer till the skins crack. Strain through cheesecloth.

ROSE-HIP JELLY

Use a cup of juice to a cup of sugar. Put 4 cups juice, 1/4 cup lemon juice, and 1 package powdered pectin into the kettle. Bring to a rolling boil. Add 4 cups sugar all at once. Return to a boil and boil till jelly sheets from spoon.

ROSE-HIP SYRUP

Follow the same recipe as above, but stop the boiling before the syrup gets to the sheeting stage—about four minutes. It should be syrupy-thick, but not watery-thin. You just have to taste-test from time to time.

Rose-hip jelly and syrup are light pinkish in color. If you want, you can dry rose hips like raisins and sprinkle on your cereals. They are chock full of vitamin C.

Having several kinds of jellies and syrups (chokecherry, rose hip, crabapple, raspberry, blackberry—whatever berry) on hand to serve guests impresses them mightily. Be modest about accepting compliments but be sure to give a jar of something to the visitor as she or he departs your abode. (Even if you can't stand the person.)

SECTION TWENTY-NINE

Surviving the Annual Christmas Program at the Schoolhouse

Wherein one learns how to cope with the planned chaos of the annual school Christmas program.

As Christmas draws nearer, you begin to feel you may be gaining on all the confusion. Then Number One son mentions the school Christmas program again. What he failed to mention till now (the day before the event) is that he has a starring role in the Christmas play. This year, that pretty, enthusiastic, new, young teacher has the kids doing a skit about Sir Galahad and the Holy Grail. This particular bit was chosen mainly because the eighth grade has a preponderance of big boys this year. They're all making their own personal swords to wave about. What they're not making is the rest of the costume. Your son needs, he says, a knighthood suit. He also says all the other mothers have already made the other kids' suits. Jokingly you ask if he needs chain mail armor or will a simple knightly tunic do. He says he needs chain mail armor. Do not panic or hit him. Calmly pick up the wire basket and go gather eggs. To the chickens, express your opinions freely. They won't care, nor will they argue. Very possibly they will cluckingly suggest something helpful.

In any case, inspiration always strikes. Ask your son to don an old pair of jeans, an old work shirt and his old cracked riding boots. Have him wrap a towel around his head to more or less mask his face. Then spray him all over silver up to his neck. (Spraying is best done outdoors). By the time you finish you'll both be giggling hysterically. Liberate some sheet tin from the shop. Cut two ankle-to-knee shin guards and attach around his lower leg with thongs. Then cut two tin, half-moon shapes big enough to act like a short apron. One to hang in front and one to hang over the rear of your silver-hued son. The chain mail could have been a problem till you noticed the fireplace screen which just happens to be a flexible metal mesh. Borrow that. When circled around son's body and pushed up under the armpits, it hangs just above the tin aprons. Attach it by leather straps or baling twine over the shoulders. Spray again. What to put on his head is a puzzler till you recall your favorite rib-knit ski mask. Mercifully you spray that silver before you put it on son's head* (see footnote). All in all, Number One son makes a fine-looking Sir Galahad.

*To make a knight requires three and a half cans of silver paint. (Do not spray the head gear with the kid in it.)

Number Two son has been turning green with envy watching his brother turn silver. Then teacher invites small son to learn a poem to recite as the opening piece of the program. Youngest son is torn between terror and a desire to emulate his big brother.

The night of the performance is the usual planned chaos. The two-room, two-teacher rural school bursts with Christmas artistry created by the kids. In the largest room (about as big as a double garage) incredibly hard benches and chairs fill all the available free space in front of the homemade plyboard stage. Seated on these torture racks are all the proud mothers and fathers holding preschoolers on their laps. Aunts, uncles, grandparents, and neighbors swell the ranks. You are seated on the aisle on the hardest bench halfway down the room where you have a direct view of the stage.

At the side of the room stands an enormous Christmas tree decorated with paper chains and foil things. On top, a giant, smiling, papier-mâché angel leans precariously sideways. One wing has come partly unglued and dangles upside down. Folding chairs line up along the side wall behind the Christmas tree. Those persons seated there have to crane around the foliage to see the stage.

As the curtain jerkily slides open on its safety-pin rings, Number Two son stands stage center frowning horribly at the audience. Hands jammed deep in the pockets of his brand-new denim jeans, he rips off two lines of "Christmas comes but once a year ..." then goes into a decline. You sit, aching with the urge to help him, when a piercing whisper by the official backstage prompter gets him going again. But not for long. For the next nine out of the ten lines left, he has to be prompted. Although he keeps his head rigidly facing the audience, his eyes slip sideways at the end of each line. Then they slide back for delivery. The last line about presents and Christmas cheer he rips off with relieved finality.

With no hesitation for any possible applause, he leaps forward off the stage, marches straight on past you, the proud mother all ready with sympathetic congratulations. Somehow, his four-year-old determination parts the sea of people till he can clamber up beside his dad who is seated on a table against the back wall* (see footnote). Although Dad extends a congratulatory pat on the shoulder, no words are exchanged, or necessary. The two men merely sit, duty done, right fists under chins, staring at the stage.

*Country men cluster like flies at the back and sides of any room filled with people.

Recitations, songs, and Christmassy skits follow the opening number. Number One son becomes a star that night. His knighthood suit is an outstanding success. He even clanks as he walks. Although he drops the Holy Grail once, he cleverly tips it upright with his sword. A sixth-grade girl, playing a distressed maiden, develops a bloody nose but manages to keep her head tilted back so it doesn't drip on her dress. At a particular moment, the script calls for her to move near the edge of the stage—at which point a hand pokes forth holding a wad of saving tissue. Her remaining lines are delivered in a slightly muffled, nasal twang. At the end of the performance, the entire cast receives a seated ovation.

In a moment all the audience begins to filter back to the refreshment table. Everybody stands around munching and talking. As the evening draws to a close, catch up with Number One son and make sure he has all his armor collected. You'll need the fireplace screen.

Finally, the refreshment table is cleared by the mothers. The older boys collect the used paper plates and cups, the girls vacuum the floor, and the teachers walk around slightly glassy-eyed. It's been an awfully long week directing all those child stars.

As the family drives home in a warm, Christmassy glow, you remember that Number One son will be home for two whole weeks to help his dad feed the cattle! Gratefully extend warm congratulations on his performance one more time. Number Two son is asleep on your lap.

SEASON'S GREETINGS

Along in December a fat jolly man
Hitched up his reindeer, count 'em—four spans!
He gathered his presents and filled up his sleigh.
He was ready to tackle his own special day.

The reindeer plumb sparkled from antlers to toes,
And Rudolph, the leader, had shined up his nose,
When out from a snow bank a bureaucrat popped,
And said, "Say there, Santa, you've got to be stopped.
"Cuz the guy who inspects all the brands in the state
Must eyeball your prancers and tell you their fate,
And those without markings or antlers on wrong
Will have to remain, they can't go along.

"They can't leave the Pole unless they've been checked
For vibrio, blackleg, and are they stiff-necked?
You must have a permit," he said with a sigh,
"If eight reindeer go flying the sky.

"And have you permission to land on the roofs
And raise such a clatter with thirty-two hoofs?
And driving a sleigh that is hitched to some deer,
You've got to admit sounds awfully queer.

"You're way out of date in that silly red suit,
And gifts from a backpack just do not compute!"
Well, Santa and Rudolph and all eight of the deer
Felt pretty bad on this night of good cheer.

Then floating on wind and drifting white snow
Came the sound of a kid who was whispering low—
"I'm waiting for Santa," said a little girl's voice,

And Santa Claus' spirits began to rejoice.

He waggled his arms and he said, "Read my lips;
I'm going," he growled with hands on his hips.
"You can keep all your permits for you, sir, are weird."
Then he pulled on his mittens and fluffed up his beard.

And laughing, he strode to his shiny, black sleigh,
"I've work I must do, and I'm on my way,
For the world is a-turning with mighty good folk."
And then with a grin, added, "Santa has spoke."

And suddenly there in the black of the night,
The stars formed a pathway of cobblestone light,
And Santa took off with a flick of his whip
And scattered the stars like the wake of a ship.

And whistling a ditty and reciting a rhyme
Santa made it to all in plenty of time
So here's a remembrance to friends near and far
To wish you glad tidings wherever you are.

SECTION THIRTY

Watching TV in Winter (If You Have TV)

Wherein one learns that vast numbers of women wear makeup, fancy dresses, and high heels EVERY BLESSED DAY. And wherein you wonder how those women would look wearing false eyelashes while feeding pigs in a blizzard.

The invention of television has been a blessing to all Country Women, especially in winter. Do not accept the notion you are being intellectually talked down to, or that your morals are being destroyed, or that your children will be seduced into violent antisocial behavior. Close association with farm animals exposes farm and ranch kids to a certain amount of sex and violence as a matter of course. The stuff on TV is often pallid by comparison to what happens nearly daily on the average ranch or farm . . . well, it used to be. Nowadays, I'm not so sure. For your husband, television is soothing relaxation and cheap entertainment after a day of battling the elements. Evening on a ranch or farm is a wonderful time of day. It's usually short on conversation as your Country Man relaxes on the couch till unconsciousness arrives. BEWARE! Do not turn off that TV! To do so will awaken husband who will claim he is watching. Do not bother to resent or take offense or feel in the slightest neglected when husband dissolves into sleep each evening. Be grateful. Remember, if he's asleep he's not pestering you while you're trying to catch up on your sewing or painting the woodwork or mending or a thousand other things. It is a basic law of nature that an idle man kept awake will pester. This takes the form of asking rhetorical questions or making comments on what YOU are doing. Or else he's teasing the dogs and cats or digging into the freezer for some more ice cream. Occasionally there are other interesting developments, but you can't always count on that.

For you as a Country Woman, TV has top priority on the list of household aids. In the evening it's company for you while your husband sleeps on the sofa. Those depressing monotonous jobs like overall patching and sock sorting can be made less screamingly boring with a TV set going. Also you can put in a lot of knitting time on Christmas presents for relatives* (see footnote).

The primary and most important function of the electronic miracle of TV is the glimpse of other lifestyles. Even the word "lifestyle" came to you via

*A hand-knitted something wrapped around a jar of homemade jelly impresses city relatives who always ask, "Did you really make this yourself?"

TV. You didn't know you had one till you heard one of those commentators say so.

But now that you know about your lifestyle, it's up to you to take advantage of what you learn. Watch with interest the kinds of duties involved in other people's careers and/or home life. As the young woman on the screen appears in daily costume change number three, you can ruminate about the rip in the paint-stained sweatshirt you've been wearing since yesterday. As you slip out of your high-laced work boots, note the panty-hosed legs of the TV model (male or female) demonstrating the toughness of hosiery fibers. Watch the model, dressed in a smart, sassy outfit, dash prettily up and down a TV pseudo-airplane aisle as you meditate on whether you should put on that same pair of jeans again to slop the hogs. Note the velvet pantsuit the TV lady is wearing as a pre-luncheon outfit. Watch the TV woman's long, slim, tapered fingers with the polished, shaped nails lift a stemmed glass. Glance at your own hands and note that the Band-Aid on your index finger where you rammed the barb wire into it needs changing.

It's OK to allow the influence of television to inspire you every so often to evolve a personal campaign of self improvement. First make a comparison study. Compare your livestock feeding garb with the elegant TV woman walking a dog. Her glimmering, pale blue suit is topped with a short, classic-line, high-collared topcoat for proper winter walking. Down the city street she strolls, expensive kid gloves grasping a leather rhinestone-embossed leash, on the end of which strains a small, fluffy canine. The lady's face is a marvel of false eyelashes, blue eyeliner, and peach complexion, no doubt achieved through the use of creams, potions, and royal jelly* (see footnote).

Think about the woman in the see-through peignoir sipping her morning mountain-grown coffee as you don your striped coveralls and raggedy denim jacket. Over your stubble-nailed fingers, draw yellow, cotton work gloves three sizes too big with a hole in each thumb, and think of Barbara Walters and Katie Couric. Arriving at the pig pens, shield your naked unmascared eyes from the elements by pulling lower the brim of your old aviator cap with the ear flaps. As you dump feed into a pig trough that has

*You tried royal jelly once. But since you couldn't afford $55.00 an ounce for the stuff, you applied a bit of honey wax scraped off the top of the table honey. After it dried, the muscles of your face did tighten up and presumably the wax cleansed the pores. There was no doubt where you were all day. Everybody merely looked for the cloud of flies.

no rhinestones, try not to giggle as you think of the lady reading the label on the dog food can through her riffle of false eyelashes.

By all means try false eyelashes, but practice wearing them long enough to strengthen your eyelid muscles. Otherwise the lashes keep forcing your eyes closed making it difficult to see. Such a handicap can lead to bruises and lacerations caused from stumbling into objects.

Go ahead and wear the lashes as far as the pig sheds. However, watch for gusts of wind which can catch you just as you scoop the ground grain into the pig troughs. The granules whip into your eyes causing lots of blinking and watering. Don't forget the false lashes as you begin pulling at your upper eyelid in order to blink out the grain dust. Otherwise you may yank off a lash and then hastily attempt to poke it back. In doing so you might spear a lash into your cornea causing terrible irritation and have to wear a black patch for a week* (see footnote).

As a true Country Woman, you can spend a lot to time laughing yourself silly over the hazards of trying to keep up with the beauties on TV. The whole family is still talking about the time somebody pasted a set of false eyelashes on Lily the sow-pig. To this day no one has admitted responsibility for "gilding the Lily."

Do not, however, despair. One day, television will discover the Country Woman audience and will slant programs accordingly. (Of course such programs won't have any basis in *real* country living.) They'll spiel off advertising slogans like:

Ladies, you'll never complain again about excrement rings around the floor if you use new "Gorp." Gorp is also a terrific restorative to aged, ugly hands.

Hair spray ads will show a neat, white-railed pig pen with a neat, white-denim trousered woman holding aloft a bottle of *Hold It Forever* hair spray. She bends and picks up a baby piglet. Her pretty locks bounce softly around her face as the piglet nuzzles her jugular vein. In a syrupy voice she compares the curl in her hair with the curl in the piglet's tail.

I use Hold It Forever just like Miss Piggy
Try it, you'll like it, your locks will be spiffy.

*Your witty husband wittily offers you a Hathaway shirt to go with the patch.

Other ads will feature a sexy, young, female beauty half naked, lying sideways on a pile of straw, beckoning to a young male who is approaching, bulgy-eyed, astride a tractor. She lisps, *Straw-Pile Perfume gets ME what I want!*

Future farm- and ranch-oriented soap commercials will portray a lady wearing a checkered gingham "farm" dress. She is standing over a laundry pile. Over the dress she wears a folksy apron with a pocket that's never seen a smashed chicken egg. Gaily she looks into the camera and smilingly proclaims:

One detergent—four settings—colors,
whites, polyester, and a special one for POOP!

Then she holds up soiled coveralls stained with a simulated brown substance.

There are those times during storms when the TV aerial your husband has established on top of the pine tree on top of the hill, blows down. Then perhaps for days or weeks the ranch exists without advice from the outside world. But when the WORD returns to you, nothing is lost. The commercials are the same. The politicians are the same and the laundry instructions remain the same. Indubitably, the farm chores remain totally *un*-simulated.

SECTION THIRTY-ONE

Waiting for Spring

Wherein one learns how to talk about hard winters, discuss how long the feed will hold out, and plan what to do to avoid winter doldrums.

As soon as the holidays have been met and conquered, the ranch enters into a period known as "Waiting for Spring." Waiting for spring includes that time from January second up to Easter. If the weather is fair to good, your man worries about shortage of moisture in the mountains and imminent terrible storms. If the weather dumps snow and cold over the land, he worries about blowing winds, drifting snow, shortage of feed, and the heifers who are due to calve in February.

"This is the worst it's been since the winter of '38!" says your man. (He wouldn't have been more than two in '38—at least in the recent century.) Grunt your appreciation of this remarkable fact as you flip over a breakfast hotcake.

As a good Country Woman, appear to take serious interest in all weather talk, even though you've heard all the comparisons before. Weather is part of your daily routine. The kids go off to catch the school bus just before the weather report. Feeding the cattle and pigs occurs right after the weather report. Adjust layers of clothing according to the predictions. On mornings below zero or when a storm is predicted, send the kids off to school with their lunch packed in a double-layer, extra-large grocery sack. In the bottom, pack extra mittens, muffler, sweater, and a stocking cap. Sometimes the kids have to tote the giant lunch sack back and forth for several weeks, but a country kid is used to toting.

In January and February, TAXES drive farm and ranch operators berserk. The ranch boasts a businesslike, hardbound, many-columned record book or—for the computer literate agricultural operator—the daggoned computer programs that claim to make it "easy"—even a "pleasure" to master tax preparation. Yeah, right.

However, it's that giant calendar on the wall next to the telephone that serves as a daily diary. In case of fire, first grab the calendar, then look for the kids. Your husband never forgets to mark down a transaction. He records births, weights, sales, hay tonnage—everything but your birthday and the egg money. Egg money income is your job. And a nice lucrative bit of extra income it is, too. Since the farm pays for the feed, you can keep all the income from egg sales as clear profit for yourself. Never reveal to the family exactly what you derive from egg sales. Ignore any person who attempts to point out that if the farm is buying the feed, then you aren't REALLY making clear profit. Just go ahead and happily spend the money.

When TAX TIME arrives, a strange thing occurs. As it becomes necessary to cope with the incredible mountain of paperwork required to allow the farm or ranch to continue existing one more year, your husband suddenly goes blind and develops a severe problem in his writing hand. He "lets" you "go ahead and do the books." He will, however, sign the forms.

During the waiting for spring months, you are likely to succumb to an affliction called obesity. To try to keep things cheery, it behooves you as a good Country Woman to bake plenty of cakes, pies, cookies, and especially doughnuts. For some reason, fresh doughnuts will cure the calving and blizzard blues faster than anything. After awhile it begins to look as if a doughnut the size of an inflated inner tube is attached around your middle. This is a problem for which no lasting solution has yet been devised.

While you and your spouse sit around, fatly watching the weather and eating doughnuts, mail-order catalogs bloom. Surrounding himself with farm, ranch, and machinery catalogs, your husband dreams of a new calf resuscitator of his very own. He leafs through page after page of tractor parts he must have. He diagrams new corrals he's going to build. He pores over catalogs advertising the Super Bull semen with which he plans to artificially inseminate his purebred heifers.

While your mate dreams through his wish books at one end of the kitchen table, you are immersed in your own fantasy world at the other* (see footnote). Your magic pile consists of seed catalogs, household furniture catalogs, book lists, and travel brochures.

In the seed catalog, make a check mark by everything you can possibly want. Then go back and put an X by those practical things you absolutely need. Go through once more and place a circle next to the items that are on the borderline between absolute need and interesting experiment. Last of all, fill in the order blank with those choices that bear all three marks. Then match the total expenditures with what you think you can afford.

*It's a country law that opening mail, kids doing homework, and catalog browsing are all done at the kitchen table regardless of any specially designated areas elsewhere.

Since you are saving your egg money for extra-special, super-interesting seeds, go out to the henhouse from time to time and add more vitamins and laying mash to the chickens' feed.

As you bite into still another doughnut, sort through the travel folders. Write letters to different agencies so they may advise you regarding group and family plans. They can't tell by the handwriting that you're a pudgy, doughnut-chomping individual. Use flowery phrases and let them think you're one of the bikini Riviera crowd. One year, an agency was actually going to send a representative, but it snowed and the authorities closed the pass for forty-eight hours, not to mention nobody got in or out of your lane for a week. But persevere. Travel planning is a harmless hobby and increases your knowledge of geography and history. Maybe next year the representative will actually get through the drifts and you'll make a new friend.

The household catalogs provide hours of fun and a modicum of heartbreak. Pick out a new couch to replace the broken-spring monster in the living room. Remember, however, if it's a choice between a new couch and a calf resuscitator, there is no choice.

If the creeks don't flood, the calves don't die, the market prices don't collapse, the grasshoppers don't eat the crops, the coyotes don't eat the lambs, a drought doesn't burn up the fields, and nobody gets bad sick, you may order brand-new wall to wall carpeting for the living room to replace the bright-colored tacky linoleum. In the meantime do as all good Country Women do. Maintain a firm line that you PREFER linoleum on the floors because it's "easier to clean."

When you come to the book lists, go wild. Choose everything that looks interesting. Pick out the craft and art and how-to books you love best. Then call your book-loving neighbors. Among you, work out a master list. With a little organization it's possible to have access to a variety of books you've always wanted to own without putting up all the money yourself. Also, you get to read all the juicy X-rated books those OTHER women order.

In among taxes, calving, and worrying about running out of feed for the cattle, the good Country Woman plans recreation and fun to relieve the winter doldrums for herself and family. During all the family-fun outings, it's you who ride herd on gear and equipment and sort out time schedules. Once in awhile get away by yourself to indulge in a special secret desire.

Due to the rubber flab weighing you down around your formerly svelte middle, your pre-spring secret desire is to be thin. So you succumb to an advertisement touting a "Health Spa." The spa offers ten sessions for practically no dollars. In fact, the spa is offering a whole day's worth of activities and classes in yoga, tai chi, weight-lifting, exercise machines—the whole nine yards—at no charge. It's a promotional deal to get people to sign on, so why not take advantage. You deserve to be pampered.

By some miracle, you manage to pack off kids and husband on an all-day outing without you. Having made an appointment by postcard with the "Flower of Womanhood Health and Happiness Spa," you are free to partake. The Flower of Womanhood Health and Happiness Spa is a glittering building in the City. It has a lavender front and gold turrets. Try not to be intimidated. Go in.

Assume your best Country Woman "nonchalant air." Pause inside the door to absorb the atmosphere. Observe the lavender and pink reception desk decorated with gold leaves and gold turrets. Behind the desk is a lavender and pink and gold skinny lady. Effusively nice, the skinny (the word "anorexia" pops into your mind) lady steps forward to greet you. She not only has a skinny body, but an elegant gold coiffure and pink fingernails. She wears a lovely lavender "doctor" coat.

Courage! Put yourself in her pink, lacquered hands. She is going to make you skinny, too. Try not to think about the gold and pink. Let her lead you and your frozen smile to the rear cubicle for a "weighing and measuring." Allow her to encircle your vast waist with her lavender tape measure. Share with her the slight shudder as she unbelievingly marks down the total on a pink score card.

After a series of interview questions, Ms. Lavender Lady dredges out your deepest secrets including the fact that your favorite sow's name is Peaches.

Lavender Lady leads you to "the course." Nothing makes you, as a Country Woman, more appreciative of country life than "the course." In a room the size of a small horse corral crouches an assortment of machines that move, grind, bump, massage, press, knead, and whimper. Each machine is adaptable to a portion of you.

Do not falter. View the assemblage of leering machines as you would a bunch of pesky critters. Snub yourself into the one with the giant rubber band and allow it to shake the sin from your body for awhile. Next, seat yourself on the thing that resembles rows of giant ball bearings turning on a spool. As the top row of bearing heave to the level of your seat, they knead your flabby person. When thoroughly kneaded, try the machine with a full-length canvass belt. It looks like the drive belt on a tractor-driven wood saw. Bravely lie down sideways and prop yourself on one elbow. Rhythmically, a padded hammer underneath the belt rotates on some sort of turning affair. The hammer mashes into the general vicinity of your waistline to "remove inches."

As the beat goes on, you begin to feel a strong yearning for home and family. Lying there like Cleopatra, note the other plump ladies draped around on the machinery. It's OK to smile in sympathy and offer a friendly word or two. However, try not to giggle. Shedding excess lard is not to be tittered at.

Lavender Lady then draws you onward and you get to sample other varieties of torture. In tai chi, you pose, stretch, and pose—like a kid game of "Freeze." At a signal or as Lavender Lady changes her own stance you hold whatever position she demonstrates. When you find yourself bent over like a croquet wicket, keep your giggles quiet.

Mushing on to yoga, you sit on a mat on the floor and you're instructed to breathe, which you'd thought you were already doing, but apparently there's breathing and then there's breathing. Inhale and exhale mightily but do not hyperventilate and pass out. Lavender Lady puts you through some interesting contortions that remind you how pretzels are made. Your legs and arms change places several times.

Your adventure in the Flower of Womanhood Spa is not over. Lavender Lady introduces you to weights. You hold heavy dumbbells while lifting them straight up, straight out sideways, over the shoulders, behind the back. It's worse than tossing hay bales; you fear your rotator cuff might rupture.

Lavender Lady leads you onward to a bunch of treadmills and bicycles and things that make you go faster than you want to. "To get your heart rate up," she intones.

After several hours, you're set free. When Lavender Lady invites you to sign on for a series of regular exercise sessions, smile brightly through the pain as you assure the Lavender Lady how wonderful you feel. Lie to her and solemnly state you will positively return next week. Maintain a thin posture as you exit.

Once in the car, relax your stomach and sit quietly for a moment. Rest your brow on the steering wheel if needed. Then drive to the "Jersey Lily Tearoom." Stagger inside. Order a large cup of coffee and an even larger chocolate-marshmallow ice-cream frappe with whipped cream and nuts. Dig in.

SECTION THIRTY-TWO
Sheep and Lambing
Wherein one learns what to do as a ewe midwife.

Some stockgrowers still carry on the old west war between cattle and sheep. Some cattlemen claim they wouldn't have a !@#$%^&*()!! sheep on the place because they stink, they ruin the grassland, and cattle won't graze where sheep run! As a sheep woman, pay no attention to any of that ancient controversy. Your small flock of woollies mix with the cows, the pigs, the turkeys, and the chickens. No one told them they weren't supposed to live cooperatively, so they get along fine. The extra income from the lambs and wool provides an occasional new pair of shoes or curtains or even new, heavy-duty coveralls. Two years ago, when prices were good and coyotes weren't as plentiful and "predator-friendly wolves" hadn't been invented, your lamb and wool crop purchased the new recliner chair that Blossom, the housecat, is so fond of.

Lambing season is planned for the end of April unless the bucks have gotten out and impregnated the ewes at a time unselected by you. In that case, you have to take the lambs as they come. As with pigs, cows, and heifers, the ranch or farm personnel have to keep a round-the-clock watch on the lambing shed. You, as one of the high-level personnel, have a prime responsibility. During lambing it's likely you may sleep in that recliner chair between trips to the sheep shed.

At some farms and ranches the custom is to have the heavy wool sheared off previous to lambing. Others only "tag" the ewes and wait to shear after the lambs are born. (To "tag" is to shear only the wool covering the ewe's faucets.)

The sheep shed or lambing shed is that long, low-roofed building next to the sheep meadow far away on the other side of the cattle corrals. Inside the shed two long rows of four-by-four (or larger depending on the size of your ewes) wooden pens march the length of the building. Each with its own gate, the cubicles are called "jugs." The effect can be compared to a one-level beehive or a one-story condominium for sheep. Fresh straw covers the floor of each little apartment. Into the jugs go the new mothers and their brand new lambs.

Some pens are larger, big enough to hold several mothers and children. After a day or so in a jug, a mother and baby can be turned into the larger pens with other pairs. A few more days and the new families are turned loose in the nearby meadows to graze. All these graduations help insure a lasting recognition and binding relationship between mothers and offspring. Otherwise, lambs and ewes tend to lose track of one another and bedlam ensues.

Every evening you and the kids bring the heavy ewes* (see footnote) inside the shed where they run loose in the long, open space not taken up by jugs. At 10:30 p.m., right after the news, garb yourself in your swell lambing outfit. A sheep-woman lambing costume differs very little from a pig-woman ensemble. Wear the standard coveralls over standard jeans and warm shirt. Don't forget your snaggedy denim jacket and the old aviator cap with the ear flaps, especially during night forays. Even in April, ears can become awfully cold if they're left hanging out. Wear four-buckle overshoes because it's bound to be muddy. Carry your extra strong flashlight and trudge to the shed. Once there, walk quietly among the flock. Pick up the "drops"** (see footnote) and "jug" the new mothers and their lambs—which means pick up any newborn lambs and carry them into one of the small four-by-four pens. Hold the lamb low so mamma will follow along, baa-ing. Once the pair is jugged, go get the jar of iodine from the old refrigerator medicine cabinet standing in the corner. Lift up the new baby by the front feet and slosh the end of the umbilical cord into the iodine. Usually mother is eager to help with the care of her lamb and gives no trouble. However, it's best to keep facing mamma while you handle her baby. Otherwise, you could find yourself butted off your dignity.

Sometimes a yearling ewe hates her offspring. Inexperienced, she is not quite sure what just dropped. But she is positive she wants no part of the thing. She sniffs, snorts, stamps her feet, and leaps to get away from the horrifying little creature when it tries to nurse. It's your job to convince the young ewe that motherhood is her role in the rich pattern of life. If she has had her lamb in the open space of the shed, it's up to you to coerce her into a jug. First, pick up the lamb and jug him. Then try to discuss the situation with the yearling. She, of course, panics and runs to the other end of the shed. Move quietly among the flock and try to get the new mother up against a wall or into a corner. If she still spooks away, you may need to resort to a sheep

*A "heavy ewe" has nothing to do with her weight. She is about to pop out lambs, quantity unknown.

**Picking up a drop has nothing at all to do with nefarious doings, unlawful schemes, littering, or manure.

hook, which is the original ten-foot pole. The curved metal hook on its end allows you to stand ten feet away and snag a sheep by a hind leg. Whichever method is used to capture the ewe, once you have her, grasp her firmly at the neck and quickly sling a leg over to straddle her, facing her head end. Hang on to her head with a hand on either side and start motivating toward the jug with the fresh lamb within. If the ewe is large and you're short-legged, you may find yourself in an unstable situation. In which case, if she hasn't already dumped you, get off her back. Grip her fiercely on the side of the head with one hand and grab on to her tail end with the other. Get a good, tight squeeze-hold and aim her for a jug—any jug. You can always bring her the new lamb. Once she takes off, you must pitter-patter rapidly alongside to keep up. You look a little like one of the clown acts in a performing circus as you zip around the shed. Another method is to grab her hind legs and propel her like a wheelbarrow. This works fairly well unless she outweighs you.

Once you have the reluctant yearling jugged, encourage her to allow her new child to take nourishment. Establish firm control by holding the ewe still while junior searches. Unfortunately, the nuzzling of the baby lamb may drive her to frenzy. She rips out of your grasp and skitters frantically around the jug while junior bleats pitifully. In order to promote a more meaningful mother-child relationship, take a piece of baling twine and tie mother's back leg to the pen side. Having to stand on her remaining three legs insures she can't kick her child away when he tries to find his dinner. Another method is to employ your pocket knife and cut a gunny sack lengthwise into two strips like long bandages. Pass one strip behind her front legs and around her body and the other strip around her body in front of the udder. Tie each strip to the side of the pen. The length of time required for a young, new mother to accept her lamb varies from a few hours to several days.

Sometimes a ewe has trouble delivering, particularly if it's her first born. If you observe a ewe in trouble, urge her into a jug. Then quietly grasp her around the neck in a hugging action. Lift her up so her front feet come off he ground and she sits back on her haunches, then gently ease her over on her side. Sit down facing her rear end, your legs spraddled on either side of the mamma. Perhaps the lamb's head is turned back or perhaps it's a breech presentation. Whatever the trouble, remove your gloves and insert your hand in the delivery aperture to ascertain the position of the forthcoming lamb.

By feel, adjust the head and legs to proper alignment. The front feet should slip out first with the nose resting on the forelegs. When the ewe next contracts, help her by pulling steadily on the baby feet. If all is well, out pops a lamb covered with a golden orange stuff. Make sure no membrane covers his nose, lift him up by the back feet, and shake him to promote coughing out phlegm. Then work the lamb's front feet back and forth to help insure proper breathing. As soon as he's breathing well, iodine the navel and poke him under mamma's nose (if she isn't already sniffing him) so she can get on with the job of licking off her baby. In just a little while, the lamb staggers to his feet, crowds into mother, and begins to nurse.

Sometimes a ewe can't raise her children. Sheep are inclined to double and triple offspring. Since a ewe has only two teats, unless she's an exceptional milker, one of the lambs may become a "bum."

For whatever reason, each spring sees several bum lambs. Some big ranches give away the bums to anyone who wants the trouble of raising them. At your place, you and the kids get to have all that fun as substitute mamas.

Pen all the bum lambs together as they come along and start feeding them four times a day. Begin with a sixteen ounce pop bottle (fitted out with a nipple) half full of milk per feeding. As the lambs grow, they don't have to be fed as frequently, but the amount increases to about two quarts of milk per day.

Feeding the bums requires a system. Your system is to assign both sons the responsibility. The boys learn beneficial self sufficiency, plus earning themselves some income when they sell the lambs and wool. A small problem lies in the fact that so often the kids are in school and it's you who must tramp to the lambing shed or the barn toting a mixed bag of bottles of milk. The instant you step over the threshold, the lambs recognize you as Mom, and baa piteously. When feeding bums, you need eight hands. As you stuff a bottle into the first two faces, the rest bleat, blat, and hop around trying to crawl into your pocket. Watch that you don't accidentally feed a greedy woolly twice. If you have too many to keep straight, take along a giant magic marker or piece of colored chalk. When you are holding a bottle in each hand with a lamb attached, slurping away, it becomes difficult to reach in your pocket for that marker. Therefore, previous to stuffing the lamb's faces, stuff the end of the marker or chalk into your own mouth. Then bend your head down over each little angel as he eats from the bottles you hold and swipe him across his nose. This method works OK except you're apt to jab too hard and drop the marker or jam it sideways in your face. Besides, it tends to become slippery with saliva if you have a lot of bums to mark.

Another method is the "knee catch." As soon as you have two lambs firmly established on a bottle each, place one bottle between your knees and grip it tightly. This frees one hand to reach in a pocket for the marker. Take care you do not attempt to walk while in the knee-catch position or you will lose bottle and marker and become hopelessly confused.

Of course, the savviest method is to purchase one of those buckets bearing a circle of nipples around the lower edge. With this gadget you can feed a batch of bums at a time. Or you could have your man drill a board with large enough holes to hold bottles, sort of like a spice rack. Nailed to the barn wall or a saw horse, it's an efficient feeder. The little woollies line up and suck away, tails wiggling happily.

As your little lamb family matures, they begin to follow you about the barnyard like a woolly parade, each one calling Maa-a-a-a!

Those little wiggly tails—the lambs are born with them, but shortly after birth, the tails are docked which means either they're cut off with a very sharp knife or they're "ringed." With a knife, you try for the gap between vertebral bones in the tail and slice through. Daub on some powdered disinfectant; there's not much bleeding. If you use "rings," they're weensy hard rubber circlets a little bigger than Cheerios. Using a pliers-type gismo to spread the ring, you slip the rubber onto the lambkin's tail close to the root, pull away the pliers and the green ring is so tight, it cuts off circulation. In about a week or so, the tail will fall off. Sometimes the dogs eat them. Sometimes the pigs do. Mostly they just rot and nurture the earth.

The same type rubber ring is used to turn a buck lamb into a wether lamb. Your husband has mentioned how he used to remove a lamb's privates with his teeth, but we won't go into that. Suffice to say, the same circulation deprivation caused by the rubber rings will take care of the matter.

NATURAL RECYCLING

Cousin Claude wore glasses dark,
While slouched upon the couch;
If he had to rise before mid-day,
He turned into a grouch.

Claude griped because he couldn't find
Vegetarian food.
Anything that cook had made,
He claimed just wouldn't do.

He wouldn't eat fresh eggs or bacon,
Or any kind of beef.
"For I'm a serious vegetarian,
And meat would give me grief!"

"You sure do have a weighty problem,"
Cousin Clarence said.
"I reckon country vittles
Are apt to make you dead."

Now, it was springtime after lambing,
And just that special time,
When little ram lambs undergo
A certain change of mind.

Clarence fetched the rubber rings
And all the proper tools,
And set them on the kitchen table,
While Claude sat on a stool.

"I'll take a bowl of those," said Claude,
"They must be good, they're green."

"Why, sure enough," Clarence drawled
And passed the lambing rings.

"These here are vegetarian cheerios,
I bought 'em just for you;
They're minty green for added flavor,
Though not real easy chewed."

Claude ate and ate and ate and ate,
Spooning up the rings;
It took him nearly half the mornin'
To finish off the things.

Those little zeros went down whole
And out Claude's other end,
They kept their shape and mint green tint
Through all internal bends.

"Oh, my" said Claude, his face gone pale,
"Is this a sign I'm dying?"
"Naw," said Clarence, "it's what cha call
Natural recycling."

PART FIVE

Woman Work and Woman Activity

"Woman Activity" includes all those community activities you do that you don't have time for but squeeze in anyhow."Woman Work" includes all those household duties and nasty maintenance chores, cleanup jobs, and eye-straining secretarial and bookkeeping tasks your man "lets" you do.

SECTION THIRTY-THREE

Sleep

Wherein one learns that sleep is a forbidden luxury.

If you live in the country, sleep may be defined as that restful commodity you're not allowed in the mornings. Country custom says you must get up before dawn in order to be ready for the day when it begins. This is vital. If the rancher or farmer isn't up, the sun is unable to tell if it should rise and shine. In the country, coping with any untoward happening, no matter how trivial, is accomplished by getting up an hour earlier. A serious problem means rising in the middle of the night.

Among city people, a belief exists that all country folk automatically pop up in the pre-dawn with a glad heart and a light step. Not true. As in the City, there are morning people and night people. However, in the country, regardless of your own physiological inner time clock, you are REQUIRED to be a morning person. It's a cruel world in the early hours for those Country Women who are NOT morning people. While husband rises to his greatest heights of oratory in the early morning, you can barely breathe. Your dearly beloved has learned to refrain from asking questions that require actual answers. He seems satisfied with nods, grunts, and an occasional moan, which is the best you can manage.

Although no real help exists anywhere, night-person Country Women have collaborated to formulate the following sleep rules.

Rule 1: If you help with the milking, never volunteer for the morning shift. When your husband rattles the milk buckets on his way out to the barn, turn over in bed, place pillow over head, and grab a few more moments of sleep.

Shameful Habit: Failure of the Country Woman to leap up and prepare hot coffee for her Country Man. Comforting thought: Nobody will hang you if you ignore that custom.

How to get around shameful habit: (a) Make a fresh pot of coffee the night before and leave it plugged in or (b) assemble a giant extension cord that runs from the kitchen, down the hall, around the corner, and into your bedroom to fetch up at the wall plug on your side of the bed. Splice into the cord one of those on-off switches. The second the alarm screams in the morning, without even removing your head from under the covers, reach out and push the switch to "on" position. By the time your man finishes dressing and firing up the heater stove, the coffee will be fresh and hot to greet him. This entire system can be constructed for under five dollars depending on how far it is to the bedroom* (see footnote). When you hear the back door slam signifying the cow-milker's return, crawl out of the nest, slide into some sort of clothes, and be standing in the kitchen holding a cup of coffee and a spatula to make it appear as though breakfast is practically ready.

Rule Two: Establish one day (usually Sunday) as your personal and deserved sleep-in morning. Refuse to crawl out before at least 9:00 a.m. Do not succumb to the myth that the farm or ranch will fly to bits without you on deck at 5:00 a.m.

Rule Three: To insure non-disturbance for your chosen morning sleep orgy, stash away a batch of yummy sugared cinnamon rolls in some secret place. At bedtime the night before (make sure everyone else is already sacked in), unearth one of those cake-saver containers. On top, leave a note, which may be reused each week. The note need not be poetic unless you are so inclined. The message directs early risers to eat all they wish QUIETLY, while warning of fatal disaster should any attempt be made to storm the sleep fortress before YOUR chosen hour.

Rule Four: Should any of the preceding rules be infringed upon, do not kill in anger. Say nothing. But the following week, put out the cake box in the center of the table again. Inside the cake cover, place a note that says,

*Do it yourself plans for the Country Woman coffee-pot-switch-on-system can be sent on request. Clear diagrams included.

"Surprise! No goodies today. Tune in again next week." (This note may also be saved for reuse.) The following Sunday, reintroduce the cinnamon rolls with the standard help-yourself note. This method is NOT a can't-fail prescription, but it works pretty well most of the time.

Early out of the nest in the mornings is such a sacred country custom that a Country Man would die before he'd be caught in bed after sun-up. So strong is this habit that Country Men who need to phone each other for business reasons always do so between five and six-thirty a.m. The stated reason is so the first Country Man can catch the other "before he goes outside." When the phone screeches at 5:30 a.m. in the pre-dawn, your husband immediately leaps from betwixt the covers, grabs a shirt, staggers to the phone and bellows "HELLO!" Your Country Man immediately claims he is up to the caller; in fact, he says, he was just on his way to the milk barn. (Meanwhile he is shrugging into his shirt, his naked legs hanging out below, gathering goose bumps.) If he needs to call a neighbor, your spouse sets his alarm an hour early so he won't be late to call neighbor Sam, because "Old Sam gets up pretty early." You know darned well Old Sam also is standing there at his end of the line half-naked and shivering. But do not offer any comment. Just press your coffee switch-on button and go back to sleep.

On a morning when your mate may feel justified in sleeping in, after perhaps being up with a critter half the night, he rises in the pre-dawn as usual, gets completely dressed, then lies down for a nap on the couch. First, however, he turns on a light in the kitchen so that any possible early passersby on the county road will know he's up and about.

SECTION THIRTY-FOUR

Coffee, Cookies, and Daily Meals

Wherein one learns that coffee and cookies are a country staple. One also learns country-meal preparation and how to develop certain shortcuts for sanity's sake.

It is of prime importance NEVER to run out of coffee on the farm or ranch. ALWAYS keep the pot hot. Country custom says gallons of the brew must be consumed daily. Farmers and ranchers have stomachs, livers, and kidneys impervious to strong beverage. NEVER substitute tea, cocoa, or some other new-fangled city drink. Even if the Country Man really doesn't like coffee, he would rather turn down coffee than tea.

To go with the coffee, cookies are an absolute must. Always have the cookie jar full—of course—of HOMEMADE cookies. If store-boughts are substituted, profuse apologies must be offered. Make great batches ahead of time and store. Country custom says the good Country Wife always piles a plate high with cookies to offer guests along with coffee. Do not ask your guests if they want any. Repeat: Do not offer a choice. Place the heaped plate in the middle of the kitchen table around which everyone is, of course, sitting.

In the country, meals must be served at given hours promptly and constantly. Lots of food must be available at all times. Vast quantities of meat, bread, and potatoes must appear three times a day. Twice a day, dessert is a requirement (and always cake or pie one of those times; can be fruit on the other time) to round off the meal. (Cookies are not generally used as dessert, but maintained as filler for mid-morning or mid-afternoon coffee and/or drop-in visitors.) In summer, it is OK to add ice cream to the dessert menu. Preferably homemade and usually glopped on top of the cake or pie.

Breakfast starts with the usual gallons of coffee followed by several eggs, pancakes, and/or hot cereal with fresh cream. If any potatoes are left from the night before, fry those also. Fruit juice is optional. If you want your Country Man to have his vitamins, place a glass of juice with vitamin pills in front of him. He will automatically drink and/or swallow pills.

The one redeeming feature of breakfast is the sameness of the menu. The routine is easy to learn. This allows you, after some practice, to make breakfast while still asleep on your feet. There are no other redeeming features.

No deviation is allowed concerning dinnertime, which is ALWAYS at high noon. Prepare large joints of meat, many potatoes, kettles of gravy, salads (especially Jell-O), vegetables, gallons of coffee, and, of course, dessert. Biscuits or hot breads or homemade loaf bread is a rigid requirement. It is an absolute country law to have bread of some sort* (see footnote).

WARNING! You may set the table and put the salad on, but NEVER, NEVER place the meat and potatoes on the table until you yourself are ready to sit down. Otherwise you will find yourself eating alone. The Country Husband is not impolite; it's just that eating is a reflex and he can't resist the food if it's placed before him. An alternate plan is to remove all the kitchen chairs to a far corner or to another room, which makes it safe to put all the food on the table ahead of time. If these precautions are not observed, the Country Man will wash up, sit down to the table, and devour food. Some Country Husbands try to restrain themselves to wait for the woman to complete last-minute tasks and sit down with him, but beware! Even those nobly controlled fellows will scoop portions of food onto their plates, mash on their potatoes, pour their coffee, and clutch knife and fork, to sit poised at the ready. The second you move to sit, he has you outdistanced, no matter how fast you chew.

The one redeeming feature of the meal is that there is no dawdling. Food is met, conquered, and disposed of in record time. Conversation is limited to breathing between bites. When the final bite occurs, a last cup of coffee poured, and a toothpick arranged in the face, then the contented husband may, and probably will, talk.

Supper is slightly better in terms of preparation. As a good Country Woman, cook enough for the noon meal so that you can have leftovers. Or chop up everything and make soup. The supper meal should be at 6:00 p.m. However, some flexibility is allowed here. You can be later without stopping the sun in its tracks, but not TOO late. The clever Country Wife mixes a toddy for her man AND one for herself, then fixes the supper while he sips companionably and talks. This is one of the warm spots of the day when your husband likes to talk. Really a pleasant time, spoiled only by the fact of one more meal to organize and clean up after.

*Baking bread and biscuits are extremely high on the list of Country Woman arts.

COUNTRY WOMAN BISCUITS

2 cups any old flour
4 teaspoons baking powder
1 teaspoon salt
2 tablespoons sugar
1/2 cup lard or whatever shortening is available
3/4 cup milk *or* 3/4 cup buttermilk, mixed with 1/4 teaspoon soda

Sift the drys into a bowl. Mix in shortening with fingers to crumbly stage. Dump in the liquid and stir with a fork to a soft dough. Don't overdo the mixing! Turn out onto lightly floured board. Roll or pat out to about 3/4 inch thick. DON'T MASH! Cut with biscuit cutter, jar lid, jelly glass, or anything round about biscuit size. Place in pan and brush tops with melted butter. Bake at 450 degrees for 10 to 12 minutes. Serve with homemade jelly, jam, or honey.

SECTION THIRTY-FIVE

Washing the Separator

Wherein one learns basic attitudes toward the milk separator.

Dealing with a separator is a matter of constantly bearing a cross. A malevolent contraption of stainless steel, the separator consists of a big, fat, round bowl large enough to hold better than a milk bucket full of milk. The bowl lets the milk down through a valve opening where it enters a gismo that encases a series of metal discs layered closely together. When turned on, the casing full of discs whirls faster than the speed of light. The milk passes through these, and magically, the milk and cream flow into separate spouts that channel into separate containers. No one really knows how the machine can tell which spout leads to cream and which to milk. But it never errs. You, however, are not so perfect.

Washing the separator is a ghastly, hideously boring, cruel albatross of a chore. One ALWAYS done by the woman. He milks—you wash. There is no way out. And there is no way to make the chore pleasant. But there are a couple of things to make it less boring. For instance, you can time how long it takes you. After a week, you will have shaved several minutes off your record. Or you can wash the monster only every other day. (But don't tell anybody and keep milk for yourself only on the alternate clean days.)

If you're not there to wash the separator, leaf by scummy individual leaf, the thing viciously mucks up its innards on purpose. Years of your life as a Country Woman are spent toiling over the evil-hearted beast. All previously refined Country Women learn unique curse words as they pick up and try to fit together all those accidentally dropped separator leaves.

If your Country Man is away for a day or two, and YOU get to do the milking, the separator job remains yours. If YOU are away for a day or so, your spouse does the milking, and the separator job remains yours. The crud simply continues to build until you get home.

A third and best alternative: Get somebody else to wash the cursed thing. Since that never happens, you are stuck forever and ever. There are no redeeming features to this chore, whatsoever.

Separator-washing can, however, be employed as a strategic weapon. If any kid balks at garden weeding or wood hauling, suggest a trade in jobs. As an alternate to washing the separator, persons of any age or sex will do almost anything.

The only relief from year-around separator washing is when one milk cow goes dry and the other hasn't come fresh yet. Actually, such a situation has occurred only twice in sixteen years, so it is not a relief you can count on.

Different Country Women develop different personal techniques for separator washing. Ideally, it's best to have two giant built-in metal tub-sinks right next to the separator in the utility room just off the back porch. Above the sinks, stainless steel metal rods extend across from wall to wall forming an open-grill work shelf. Upon this, place the dripping, scalded separator parts as you interminably wash them. The big sinks can hold the entire pile of separator pieces or a whole milk bucket can be immersed.

Some separator washers prefer to charge right into the job first thing in the morning. Others, with weak stomachs and poor self discipline, are inclined to wait until the evening newscast or the Leno or Letterman shows before tackling the thing. A little comedy helps.

Many, many Country-Woman legends regarding correct separator-washing techniques have developed down through the years. Do not be intimidated. Ghosts of Separator Washers Past will not rise up to strike you down in the dishwater if you change the rules.

Once the beast is washed and hung up to dry, the nasty portion of the chore is finished. The merely irritating part comes next. Previous to the next milking, the separator has to be put together again in a certain sequence. Each part is important and each part knows where it should go even if you don't. Get one disc out of line, forget the rubber ring, or fail to tighten the cream screw, and a veritable disaster follows. When foaming milk is poured into an improperly assembled separator, the liquid immediately spurts and cascades seven feet in seven directions. Should such a horrendous catastrophe occur, think fast. Close the valve and flip the switch to off! (It is optional to employ swear words you've heard the men use in the corrals.) Then find a mop. While mopping, it's therapeutic to rehearse what you're going to say to anyone who chances upon you while sopping up the mess. (Especially Mrs. Super Rancher.) It's OK to accuse the person of deliberate sabotage. Should that person have the gall to snicker at poor you, scathingly instruct him or her where he or she can put the separator AND the milk cows.

As a last ploy, silently hand the mop over and leave. (Especially to Mrs. Super Rancher.) For days afterwards, the utility room has an odor reminiscent of un-housebroken cats. Don't let this bother you. Place two or three incense burners at strategic spots to catch the cross drafts. Light different flavors of incense sticks as burnt offerings. A few incantations can't hurt. Double the incense arsenal if Mrs. Super Rancher is expected.

SECTION THIRTY-SIX

Visitors

Wherein one learns the proper rituals, customs, and etiquette when visiting or being visited. One also learns certain techniques to divest oneself of visitors while remaining polite.

A ranch or farm may seem isolated to the casual city observer. Except when snowed in, however, there's really a lot of traffic flowing in and out the back door of any country dwelling. (In the country, front doors practically fossilize. In summer, the front door may be opened to obtain a cross breeze, but no one uses it to gain entrance.) Through the back door and into the kitchen arrive all visitors. These can be classified into two general types.

Type A: The neighboring farmer or rancher on an errand "stopping by" to see your husband. Instantly offer coffee whether you know the person or not. He will say, "No thanks," and will then wait in his pickup for your man's return from the South 40. Or he will head out for the South 40 to find your spouse. In either case, if for some reason the Country Man's "woman" is along, she will have been left sitting in the pickup. Always look out the window to see if this is so. Tell the visitor to invite his wife in, or if he's wandered off on foot after your husband to the South 40, go out to the pickup and invite her in yourself.

If your husband is home when the caller stops by, follow the same routine with slight variations. Place coffee and cookies in front of husband and caller. (First, invite them to sit down at the kitchen table.) Instruct caller to invite the wife in (you have already looked out window and observed woman in pickup). If caller says, "No thanks, I'll just be a minute," then you are obliged to go out to pickup and chat with the wife through the window. Do not put on a coat. Cross arms and shiver and chat. You do not introduce yourself and she does not introduce herself. If her husband does not emerge from the house shortly, re-invite her in. This time, she will dismount and accompany you to the house where you place coffee and pass cookies. (About then you can announce who you are and she can reciprocate with her identifying name.) When the two husbands are finished talking business, say goodbye to the woman, always ending the conversation with, "Stop by sometime." Saying goodbye to the husband is optional. As soon as the pickup pulls out of the yard, ask your husband the wife's name (this is a test). He will not know, but does know she milks eight cows by hand every day without help.

Type B visitors are social callers. These usually come in packets of two, a man and a wife. Occasionally there's a foursome. They usually have just "stopped by" and usually without phoning to see if you're home, ready for them, or have other engagements.

When you hear the vehicle enter the yard, and you see who it is and become aware you're going to be VISITED, immediately dash for the bedroom and jerk up the covers on the unmade bed. Leap back to the kitchen. Wipe off table. Go to door carrying dishtowel. Smile broadly and joyfully.

When everyone is seated at the kitchen table, spread around the coffee cups and break out the cookies. Make some attempt to coax everyone to come sit in the living room. You may or may not be successful. Usually the men retire to the comfort of the living room armchairs, leaving the women hauling the coffee and cookies after them. Then the women return to sit on hard chairs in the kitchen, occasionally interrupting their chatter to pack in the coffee pot to the living room and refill the men's cups. Sometimes the order is reversed and the women sit in the living room, leaving the men in the kitchen. However, the women still see to it that the men's cups are kept filled.

Country custom says that men and women shall not sit in the same room and converse TOGETHER, especially if more than one other couple is present. If only one couple is visiting, women sometimes may sit in the same room with men, but are expected to converse with each other—not with the men.

When the situations are reversed and you do the visiting, follow the same customs. Visiting can also be classified into two general categories.

Type A Visiting includes accompanying your husband on an errand. At times, you "ride along" while your man "stops by" another ranch or farm on a matter of business with the other Country Man. You will be left sitting in the pickup talking through the window to the other arm-crossed shivering Country Wife because your husband will "just be a minute." Often, however, there is no wife present to talk with. Country Men prefer to track each other down in unusual and generally removed places—like halfway down the irrigation ditch. For these lengthy moments of waiting, the wise Country Woman keeps a care package in the glove compartment of the pickup. Frequently a sanity saver, the package contains chewing gum, supplementary food, tissue, comb, lipstick, and MOST important—a novel, magazine, or crossword-puzzle book to read during those "few minutes" of waiting.

Sometimes your man and the other Country Man conduct their business through the pickup window. You still can manage to get in a lot of reading as a great many moments will elapse between remarks, none of which will be directed to you anyway. Ten to fifteen minutes or a whole chapter pulsates by before the point of the visit slips into the conversation. The weather, the calves, the sheep, the neighbor's lame colt, the price of cattle and hogs, ditch cleaning, another neighbor's drunken hired hand—all these will be mentioned while the conversationalists gaze at the far horizon. In fact, the two men will say their goodbyes, your husband will start the pickup motor, you will close your book, and ready your goodbye smile. That's when your man off-handedly inquires, "By the way . . ." and reveals the main reason he's "stopped by." At this point, you can get in another half chapter.

Always it is a pleasure to receive visitors and catch up on what's going on in the county. Sometimes, however, you may be driven to the point of rudeness when certain thoughtless visitors begin to camp underfoot while your lambs go unfed and the separator remains unwashed. There's always a family like Pudwell Buford and his brood of six kids who visit en masse. Pudwell and his wife, Luella, arrive at 8:15 in the morning and settle themselves in. Mr. Buford awaits your husband's return from the feed ground, he says. Luella just likes to "come along and visit." The kids range in unwashed ages from two to thirteen. Pudwell believes that a family that lives together ought to visit together.

To rid yourself of Pudwell Buford–type visitors, be bold. Country custom says you must not point to the door and order them out. Country custom also says if they're still around at meal time, you must feed them. Maintain cheer, pour coffee once around, and put out a skimpy plate of cookies. When the cookies are gone and the coffee pot has gone dry, do not replenish. If it is summer, close all the doors and windows, put on a sweater, and mention you have a chill. Then build a roaring fire in the cookstove. Cough a lot and sniffle some. When the Bufords make noises about leaving, give them a dozen eggs on the way out.

If it's wintertime when the Buford tribe comes calling, do not falter. Open all the doors and windows to "air the rooms." You can wear something especially warm and stand over the cookstove. Let the coffee pot run dry and wait for their chilblains to raise. Give them a dozen eggs on the way out.

SHEFFIELD
FENCE

SECTION THIRTY-SEVEN

Coping with the Annual 4-H Fair

Wherein one learns to bear up nobly.

Each year the county sponsors a 4-H Fair where kids can exhibit animals and display produce they've raised themselves. Boys and girls from seven through high school are eligible. To be a 4-H-er, one strives for a *Head That Is Clear, Hands That Are Useful, A Heart That Is Kind, and Health That Is Good.* Crafts, needlework, sewing, woodworking, and scrapbook projects dealing with farm and range management may also be submitted. Vegetables and crop plants are harvested at peak perfection and presented. Cookery, from biscuits to pies and cakes and even full meals are entered. Selected judges taste these offerings* (see footnote).

For the parents of a 4-H child, the experience is not only inspiring, it's also a test of patience, agility, and mental balance. As a Country Kid's Mother, you automatically participate in 4-H activities as chief honcho, herder, binder-up of wounds, salver of hurt feelings, and fierce protector of your own offspring.

Last year, Number One son entered his registered Hereford bull calf in the competition. He had purchased the calf with his own money (pretty much). By Fair Time, the little darling had grown to eight hundred pounds of stubborn muscle and bone (the calf, not the kid). All spring and summer, "Colorado's Golden Boy" (the name listed on his papers) had been a constant companion to Number One son, who for some reason, decided to call the calf "Blister."

Early in life Blister must be taught to lead. When you pass by the corral one morning and observe a calf galloping and bucking at the end of a halter rope, ask no questions. At the other end is a person you recognize as your son, heels dug into the turf, body slanted backwards, grimacing ferociously while he hollers, "Whoa!" and other words you didn't realize he ever employed. A determined lurch by the rambunctious calf jerks the 110-pound boy off his feet to land belly down, still hanging onto his end of the halter rope. Do not panic. Quietly go on about your business. Nobody likes to be happened upon in such an inglorious position, especially by a relative.

But there are times when you are allowed to help educate Blister. "Setting up" a calf is the art of showmanship. The animal must be taught to

*Never volunteer to be a taster.

stand with legs squarely under him and head held alertly straight ahead. A long, slim "show stick" (wrapped in braided or pseudo leather with a tasseled tip) is used in "setting up" the animal by tapping wayward feet until he stands square. The stick is also used to scratch the calf's belly, which theoretically soothes the beast and keeps him calm. Blister loves to have his belly scratched. Or he does now. At first, while your son made repeated attempts to "set up" Blister, you were chief belly scratcher. Blister, however, often failed to realize he was enjoying himself and kicked the stick plumb out of your hand.

Blister must be bathed, combed, brushed, "blowed," and clipped. His horns must be sanded and oiled and his toenails trimmed. The whole family gathers to hold things while Number One son "pretties up Blister." Garbed in a yellow rain slicker and four-buckle overshoes, your son takes the hose and turns a stream of water onto Blister. Then a machine that looks like a vacuum cleaner is employed to blow and fluff him dry. As a matter of fact, the machine probably IS a vacuum cleaner—yours. After blowing and fluffing, Number One son, aided by four-and-a-half-year-old Number Two son, goes at Blister with brush and comb. At the front end, your husband gives advice about sanding Blister's horns and applying oil to make them shine.

At the back end hangs Blister's tail, an extremity which requires a particularly deft touch. Since you are the deft-touch expert of the family, this job is assigned to you. First, clip the tail from its root end to just above the fluffy part called "the switch." WARNING: The tail of a bull hangs limply downward. Unlike a dog's, a bull's tail doesn't wag. However, it can suddenly whip up, around, and back and forth like an aggressive snake, which could assault you on face or body. Should the tail chance to be untidy, you may receive more than a mere lashing. Therefore, it is best to tie the tip of the tail to one of the bull's legs. (A back one, of course.) Starting at the base, clip carefully downwards, all around, to just above the switch. Then fill a bucket with warm soapy water and dip Blister's switch into it. (The warm water is for your benefit, not the bull's.) Slip on those big, black or orange rubber work gloves, pick up a stiff scrub brush, and proceed to shampoo Blister's switch. Chances are, what with everything else that's taking place all over him, he won't mind you messing about with his tail. After the shampoo, brush, comb, and fluff the switch* (see footnote). Caution: While you are beautifying Blister's tail, he may, from time to time, rid himself of waste

*Almost always, switch grooming reminds you to make an appointment at the local people beauty emporium to have your hair done.

material. Try not to let his freshly shampooed switch become tainted. To keep the switch pristine you can tie a plastic sack or an old pillowcase around it.

Finally, Fair Time arrives. Blister is transported along with a highly nervous son to the fairgrounds where the calf is tied in a stall heaped with fresh clean straw. Until the moment the judges call for him, young son constantly brushes, combs, and fluffs Blister. Then the MOMENT arrives. You watch proudly as Number One son, duded up in white western shirt AND tie, pearl grey Stetson, brand new pants, and much-polished boots, leads Blister into the ring where he competes against the other entrants. Eight kids, ranging in size from plumb puny to medium to tallish, stand big-eyed before the almighty judge. The critters stand there, too, even bigger-eyed and more nervous than their young masters and mistresses. The judge considers each animal carefully. How it stands, its conformation, its general bearing, its grooming and cleanliness, and how well the exhibitor manages the critter are all counted into the final score. Every show person in the ring is constantly combing, brushing, and scratching bellies while praying his or her animal will remain CALM. A great deal hinges on cleanliness and calmliness.

At all times, the 4-H exhibitor must keep his animal between himself and the judge. Eyes glued on the judge, the kids comb, brush, and scratch bellies as they set up their charges. You stand at the sidelines, fingers crossed as your son turns Blister in a tight circle. Last year his steer calf decided enough was enough and commenced to gallop toward the nearest exit, dragging your offspring along the end of the lead-rope. It had proved embarrassing when you instinctively reacted by leaping in front of the departing pair. The decamping steer instantly planted his feet, spun around and knocked over the judge who had kindly come to help.

But this year all goes well. When the winners are announced and your son walks away with a blue or possibly even a purple "best of show" ribbon, you have to slip under the grandstand for a second, because for some foolish reason your eyes are watering.

The hard work preparing the animal and the labor of special record keeping—what feeds were used, pounds gained, and rates of gain—have all been faithfully attended to by your son. Now the beautiful calf that has been so lovingly cared for must be sold. The last night of the Fair is auction sale night. Calves, pigs, sheep, rabbits, chickens, vegetables—all are auctioned off to the highest bidder. Usually local businessmen buy the meat animals at premium prices. For your son, the trauma of parting with Blister is not as severe as for some kids. Blister is sold as breeding stock, so at least he won't be slaughtered.

After the sale, Number One son's bank account swells by several hundred dollars. Already he is planning for next year when he says he's going to enter a different animal in each class. You, for one, can hardly wait.

In addition to the various stock exhibits, horsemanship competitions attract every juvenile in the county who owns a riding animal. Judging is based on performance of both horse and rider. Even little kids can enter the horsemanship event. Children under five manage the reins of their steeds, but MOTHER provides guidance by leading the pony with a halter rope. Mother leads the mount at a walk and then at a trot while the judge observes the child's horsemanship ability. No loping is required in this event. The horse and the child are willing. It's Mother who limits the action. Some pudgy mothers can barely manage to trot alongside a steed, let alone gallop.

However, the competition is keen. All the kids, ponies, horses, and mothers appear clad in assorted eye-appealing garb. At least most do. According to the rules the judge makes his choice on the basis of horsemanship and ability. It says so in the rule book. WARNING: Beware of false belief in the printed word.

At the fairgrounds, watch proudly as Number Two son, age four-and-a-half, mounts his noble steed by pulling himself up hand over hand on the stirrup leathers. Dressed in his brand-new denim pants, his almost-new straw cowboy hat, red checkered shirt, and lovingly polished boots, young son shines among all the others. How could a judge fail to be impressed with so obviously smart and handsome a child seated royally astride his recently cleaned saddle on Boomer, his very own fat, brown, half-Welsh pony?

Overwhelmed with pride, distract yourself by peering misty-eyed around the grounds at the less-fortunate mothers. Suddenly you notice Mrs. Super Rancher leading a prancing, slim, high-headed sorrel mare. A silver bedecked saddle glints, hurting your eyes. Fancy pink tassels dangle from the bridle. A matching pink leather halter and pink lead rope complete the horse's wardrobe. You consider the plain rope halter leading to your son's pony a much more practical choice.

Then your lip freezes into a permanent curl as you note Mrs. Super Rancher's five-year-old daughter, Melanie Ann. The dear little thing wears an ensemble of pink topped with a pink western hat trimmed with a silver band. White western boots match the white kid gloves encasing the tot's dainty hands. All that pink is about to give you an eye disease. Particularly as Mrs. Super Rancher wears a pink outfit that exactly matches her daughter's! You are dressed less ostentatiously in serviceable, but clean, denim jeans and plain, unadorned, western shirt. (Some people have more restrained taste in clothes than others.)

Naturally, as the judge calls for the participants, you and son are placed in line directly behind Mrs. Super Rancher and Melanie Ann. Take a deep breath and straighten yourself until you feel thin. Remember to exhale, too, so you don't black out. Take heart from your son who sits tall in the saddle. Boomer thinks the whole thing is a bore, stretches his neck and yawns widely. The motion yanks you off your brave stance.

But you can't let your son down, so lift up your chin and stride briskly forth in an attempt to wake up Boomer. Smile as you walk and trot past the judges. On the far side of the arena, bring Boomer to a halt. You are supposed to demonstrate the pony's backing ability. Glancing at your son, you notice that excitement has caused nose leakage. Whisper to him to use the tissue you sneakily hand him—rather than his sleeve. He employs the tissue and hands it back to you—of course.

Young son is thrilled with everything. Preening, he waves grandly to his friends in the audience. At that moment, Boomer decides to shake off his depression and does so with a head-to-tail quiver that rattles the saddle and the boy. Turning back from his waving, Number Two son gruffly reprimands his mount with a loud, "Cut that out, you old crowbait!" His four-and-a-half-year-old voice cuts across the crowd, clear as a bell. Boomer, taking the command as a cue to move forward, does so and treads onto your foot. Your quick-acting son pulls Boomer to a halt while he is still planted on your foot. Try not to allow tears to form. It interferes with vision and makes you seem unsportsmanlike.

When the ribbons are awarded, a blue one goes to Melanie Ann Super Rancher on her horse, "Little Misty's Golden Princess." Second and third place go to a couple of kids other than your child. Your son, on chubby "Boomer's Lament," rides proudly forward along with the remaining field of sixteen kids who each receive a plain white ribbon for "showmanship."

By this time your stomach muscles have gone slack. Concentrate everything on the pleasant smile you bestow upon Mrs. Super Rancher. Congratulate Melanie Ann and try not to lose the smile as Mrs. Super Rancher mentions what a "terribly cute little outfit" your son is wearing.

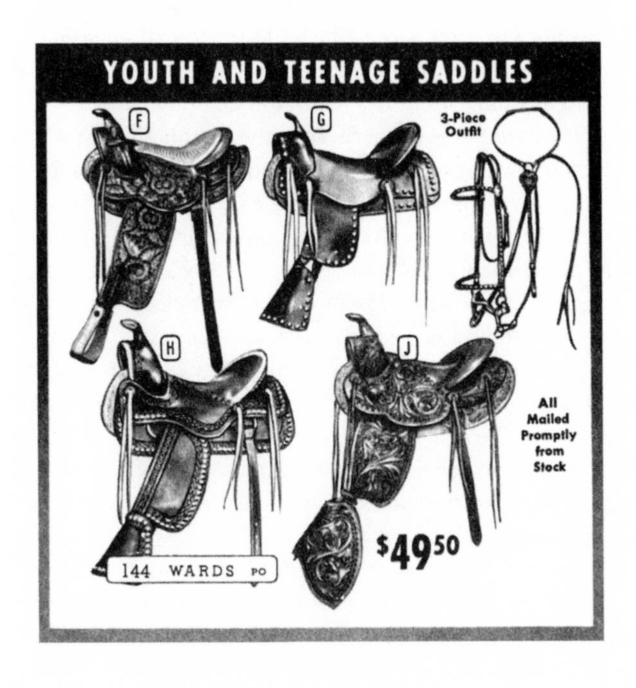

SECTION THIRTY-EIGHT

Yard-Sale-ing—An Addiction or a Pleasant Hobby?

Wherein one learns the art of outwitting other zealous Yard Salers to take the prize—be it brass ring, brass monkey, or brass bedstead.

When you've been on night duty checking heifers or sheep for a month, you begin to wonder how to reward yourself. You need a break, so you mull over options. Fly to Tahiti? Bask in Hawaii? Get a massage from a handsome hunk at an expensive spa?" (There are no handsome hunk masseuses at the spa, but you can dream, can't you?)

Then you read the weekly paper and note there's a *yard sale* this coming weekend. It's mid March and nobody holds yard sales in cold, windy weather. But it turns out Mr. and Mrs. Bifurcation are moving and want to clear out some of their stuff. If there's one thing you like better than almost anything, it's yard-sale-ing. You determine to go.

Your last heifer check is at seven in the morning, which will give you time to get to town by eight o'clock if nothing's going on in the delivery department. The advertisement said no early sales, but you want to be there at least by seven forty-five so you can be first in line. You know full well how pushy some women are, especially Mrs. Super Rancher.

If there was such a thing as Yard Sale Olympics, you would take the gold every time. You've furnished practically your entire house in bric-a-brac and antique gems from other people's cast-offs. You're not alone. In any small town or rural area, trash for some is treasure for others. During yard-sale season, the stuff levitates a few feet off the ground and makes a slow revolution and settles in the next household. Next spring, another levitation, another revolution. Some of the items have lived for a time in half the houses in the county.

Your spouse, recognizing the fever, merely says, "Never mind, I'll check the heifers. Happy hunting." You kiss him goodbye and fly to your pickup.

You put pedal to the metal and by ten minutes to eight you are eyeballing an array of stuff—furniture, dishes, knick-knacks, books, old trunks, tools—all the paraphernalia that households accumulate like fungus on cheese. Already a crowd of buyers have gathered, lingering and looking as they await the official start of sales.

Your eye falls upon an item that makes you gasp, even tremble. A two-foot-tall crockery vessel stands at the end of a line-up of goodies displayed right on the ground. You've spotted a Red Wing butter churn!

You feel as if that churn is the find of your entire yard-sale life. You swoop toward it.

So does Mrs. Super Rancher. It's still five minutes to eight. You sidle closer to the treasure. So does Mrs. Super Rancher.

"I was here first," she growls.

"No early sales," you growl back, "you have to wait for eight o'clock." With a Cheshire smile, you plant a foot on either side of the butter churn. The time is now seven fifty-seven.

Mrs. Super Rancher hisses. She zips and darts around like an angry bee. Like a sturdy oak you remain planted. At the stroke of eight, you bend, clasp the crockery butter churn in both arms and tote it to the pay desk. You don't care how much it costs—but this is a yard sale—nothing costs a whole lot.

Cradling the prize to your bosom, you stride amongst the other shoppers toward your pickup. The crowd makes way. Three women and one man applaud.

You, Her Majesty the Yard-Sale Queen, give them all a gracious smile.

SECTION THIRTY-NINE

Sidekick Savvy

Wherein one learns bits of savvy from the Country Woman's wry-eyed perspective. And one learns that incoming savvy never stops so keep a notebook handy.

The Country Woman's Credo: *The Stuff of Life Is Everywhere.*

Many years ago, in secret enclaves, Country Women began writing down pithy questions and answers about STUFF. These observations have been collected from other Country Women, discovered scratched on the walls of cave-woman dwellings, found buried in chicken coops and scribbled in Latin on duck eggs.

COUNTRY WOMAN STUFF

Wherein reading a romance novel is not the answer, but it might help. . . .

Question: What does a Country Woman do with a Country Man's underwear after he's spent a long day in the saddle?

Answer: Burn it.

Question: What do Country Women call intense physical discomfort?

Answer: Pain.

Question: What do country men call the same intense physical discomfort?

Answer: They ignore it—unless the bleeding won't stop. Reason: Testosterone overflow.

Question: Of the following, which would cause a Country Man to contemplate homicide of his mate?

1) She sets the kitchen on fire.

2) She leaves the gate open and the heifers get in with the bull.

3) She burns all the old feedstore caps, worn out boots, and frayed-to-gauze chore jackets.

Answer: Number three. Country men become emotionally attached to their STUFF.

Question: Why does a Country Woman sometimes compare her spouse to a dog?

Answer: Both growl if they think someone is bothering the pickup.

Question: What makes the spouse different from the dog?

Answer: The dog doesn't drive and he *likes* to ride in the bed of the pickup.

Question: What always occurs when the Country Woman has two bum lambs in cardboard boxes in the living room, a baby pig behind the stove, and no dishes done for three days?

Answer: Mrs. Super Country Woman calls wearing garments that cost more than your annual land payment.

Question: If the Country Woman sees Mrs. Super Country Woman or the minister approaching her door, what should she do with all the unwashed suck bottles, nipples, and vaccine guns cluttering the drain board?

Answer: Throw the stuff under the sink, in the oven, and behind the couch.

Question: Why is the Country Woman glad when Avon Lady calls?

Answer: She sells bright red lipstick and you've lost the marker you've been using on baby pigs and lambs.

Question: On farm or ranch what vicious equipment deliberately hurts, harms, or damages the Country Woman?

Answer: Farm Machinery.

Question: How does the Country Woman get out of milking the family cow?

Answer: She develops amnesia, arthritis, and impassioned animadversion for teats-pulling.

Question: On a wheat farm, what does the Country Woman do for a hobby?

Answer: Makes decorative doo-dads to sell at the Christmas bazaar.

Question: What does she do with the money earned?

Answer: That's her business.

Question: What should the Country Woman do when a male caller treats her like woodwork?

Answer: Serve mocha coffee using ex-lax flavoring.

Question: When the Country Woman is dolled up in raggedy chore clothes held together with safety pins, overshoes fastened with bale string, and she's wearing gloves divinely holey, what accessory is too much?

Answer: Pearls.

Question: What does a Country Woman get to do on branding day?

Answer: Cook up a storm for thirty or forty worthless friends of her husband.

Question: Why do so many country kids have birthdays in the same week or month?

Answer: Because, in the country, children are not allowed to be born during lambing, calving, farrowing, haying, harvesting, or shipping. Some women go eleven months before they foal.

Question: Why does a Country Woman always carry a stout hardwood stick or an old harness hame fitted with a length of strong wire in the pickup and town car?

Answer: To open those dad-blamed too-tight wire gates.

Question: Why does the Country Woman paint the above "cheater stick" pink?

Answer: So the country males will think it's a craft project and leave it alone.

Question: What is the height of frustration for a sheep-raiser woman eight months pregnant?

Answer: Getting stuck in a lambing jug and no one hears her yelp for help for three hours.

Question: What did the Country Woman say on her birthday when she opened her gift and discovered it was a new ob. strap?* (See footnote.)
Answer: Just what I need!

Question: How does the Country Woman know her spouse has used boiling water to thaw out the colostrum in the calving shed or a piece of machinery housed in the cow barn?
Answer: When she turns on the burner under the teakettle and a cloud of cowpat fog wafts from the stove.

Question: What should the Country Woman do when it's time to ride for cattle?
Answer: Saddle her own horse first. *Then* fix breakfast.

Question: What vital piece of equipment should the Country Woman never forget when riding for cattle?
Answer: Her bra.

Question: When should a Country Woman run for the hills?
Answer: When her spouse comes in the door murmuring those three little words, "Are you busy?"

Question: What does a Country Woman say when criticized by her neighbor, Mrs. Neatnik?
Answer: "At my house, the chaos is planned."

*Ob. strap: An obstetrical strap is a 3/8 to 1/2 inch wide strap about twenty inches long with metal rings on each end. You make a loop with the strap and rings and place over emerging calf's little legs. Then you pull like mad, using your arms and back or a calf puller, a fence stretcher, the pickup, or a passing elephant. Remember to go with the mama cow's contractions, not against them.

Question: What is the fastest way for the heifers to get out and into the wrong pasture, especially if you're home alone?

Answer: Put on your town-going clothes and high heels.

Question: What does a lambing shed or farrowing barn operated by a woman have that a ditto shed run by a man does not?

Answer: Electric blanket, a cot, hot chocolate, and a reading light.

COUNTRY MAN STUFF

Wherein not much is learned that we don't already know. . . .

Question: How cold does it have to be before a Country Man will lower the earflaps on his cap?

Answer: Thirty below. Unless the wife is watching. In which case he waits till his earlobes crack.

Question: What do country men (married) call a fun night out?

Answer: Staying home while the wife goes to Club, but only if she leaves a casserole in the oven.

Question: What do cowboys find rewarding?

Answer: An eight-second spine-jarring ride on a wild-eyed bronc in a rodeo arena, especially if it results in bodily damage and applause.

Question: Does a Country Man know the meaning of the word "taciturn?"

Answer: He won't say.

Question: What happens to a Country Man immediately after he carries his beloved across the threshold on his wedding day?

Answer: That's the last thing he ever again carries within the house.

Question: What happens to a Country Man's eyesight and memory immediately after marriage?

Answer: Whenever he's indoors, he goes blind and develops serious amnesia. He can't recognize socks in drawers and needs a map to locate the kitchen stove, which doesn't help as he's forgotten what a stove looks like.

Question: How could the Country Man strengthen his character when he and the wife are out in below-zero temperatures flaking hay to the cattle off the back of a pickup?

Answer: Let the wife drive.

Question: Why do country men lean on the back end of a pickup and stare down into the pickup bed?

Answer: Testosterone tea parties.

Question: Why is it that country men can drive tractors, fix mystery machinery, capture raging critters, but can't find the on/off switch on the kitchen stove?

Answer: Testosterone overflow. Again.

Question: How early in the morning does the light go on in a country kitchen?

Answer: Approximately 4 a.m.

Question: Why so early?

Answer: The Country Man is scared that early bird passers-by might notice a dark house and think the Country Man is lazing in bed.

Question: What will a rancher or farmer claim if somebody telephones him before six a.m.?

Answer: That he's been up for hours, done the feeding, milked the cows, and hauled wood.

Question: If the caller could see through the telephone line, what would be visible?

Answer: A guy with bare, hairy legs, shivering.

GENERAL STUFF

Wherein you may seek, you may search, but it's still up to you. . . .

Question: What is a windmill?

Answer: A thing with whirling arms that pumps well water, but stops dead just as the cattle come to drink.

Question: What do Government Ag subsidies do?

Answer: Support the rich and help the small farmer and rancher learn to fill out forms.

Question: In the country, what is the solution to most any problem?

Answer: Get up an hour earlier.

Question: What can the Country Man count on in livestock production?

Answer: High prices until it's his time to ship.

Question: What is the most lucrative medical practice in a rural community?

Answer: Chiropractor.

Question: What does the Country Woman do with the baby's playpen?

Answer: Put the bum lambs or piglets into it.

Question: Why do radical anti-agriculture groups wear headbands?

Answer: To prevent brain leakage.

Question: What do radical anti-agriculture groups have in common with parasites?

Answer: Eventually they destroy their hosts.

Question: Which half of the human race considers calf gonads a delicacy?

Answer: Need you ask?

Question: What do country folk call government forms?

Answer: The sluff of life.

Question: What is the most popular domestic animal raised on government grants?

Answer: Duck-billed politicians.

MORE STUFF

Wherein you learn you can't plan a whole lot, but in a hundred years it won't matter—and it doesn't matter much now. . . .

Question: Where does a rancher or a farmer take his girl on a date?

Answer: Out to watch alfalfa grow.

Question: Who is invited to a country wedding?

Answer: Everybody.

Question: Where does a newlywed country couple go for a honeymoon?

Answer: They don't. They have to stay home because it's time to calve the heifer, farrow the sows, or lamb the ewes. But, their parents get to take a trip.

Question: When do newlywed country couples *finally* get to go on a honeymoon?

Answer: Maybe after the first five years or five thousand miles on the tractor or when their own kids get married.

Question: What does the Country Woman say to a spouse who hints she's putting on weight?

Answer: "Dear, think of me as an easy keeper."

Question: What is the Country-Woman term for holding things while her Country Man is fixing stuff?

Answer: Duty palls.

Question: What bivalve seldom grows pearls?
Answer: Mountain oysters.

Question: Dressed in your Sunday best, what should you *never* do?
Answer: Chase a pig.

Question: What comes in a green can and is called hand lotion on farm or country?
Answer: Bag Balm

Question: What makes a milk cow testy?
Answer: Fingernails long enough to disc the South 40.

Question: What is wild and woolly and full of fleas?
Answer: The barn cats.

Question: What is the most valuable animal on the country or farm?
Answer: Jethro the Gelding, who never moves a muscle when kids crawl all over him.

Question: How does a Country Man keep coyotes and wolves healthy?
Answer: Raise more sheep.

Question: What do vegetarians believe?
Answer: That they will live forever if they eat only leeks and lettuce.

Question: What will vegetarians feel really silly about someday?
Answer: Lying in a hospital bed dying of nothing.

Question: What kind of animals do carnivores eat?
Answer: Vegetarians.

SECTION FORTY

Ordering a Custom-Built Country Man

Wherein one learns that when those tough times must be endured, a rich fantasy life can help.

There are times in a Country Woman's life when she wonders what her life might be like had she pursued her career in nursing or accounting or hairdressing or marine biology or sky-diving. These speculations usually occur when under duress—such as facing a rattlesnake when irrigating.

Imagination is especially stimulated on a below-zero winter day when you wake to discover three feet of snow have trapped some cattle. You as the Instant Hired Hand must don a zillion layers of clothing till you look like a short, chubby, alien being. Trudging afoot through the drifts to bring the bovines to the home place so they can be fed can lead you to think of rewards for your effort.

As you slog through heavy snow, your nose running, the tips of your fingers beginning to freeze, a cow dodges down into a gully. Your mate shouts in that elephant-calling voice that you should block the critter's path. You do, but in the process, your feet slide out from under you, which plonks your derriere deep into a drift. Your mate, a man of otherwise intrepid qualities, laughs.

You can't help but giggle too, but still . . . that's when you begin to speculate what it would be like if you could order your own customized Country Man to be delivered to your door at your convenience. You visualize that you are pushing the buttons on your touch-tone phone, wait a moment, and a velvet-tone, deep voice that reminds you of prairies greening up in spring purrs, "Thank you for calling Custom Country Man Services. For information regarding your Custom Country Man preferences," continues the voice, "press one."

You press. The phone goes, "Beep."

"Thank you," says the Voice causing your heart strings to start throbbing, "you have reached the Custom Country Man Preferred Types Department. Please choose from the following options: If you prefer a Country Man built like a little bull with muscles everywhere and of medium height, press two. If you prefer a Country Man of lanky build with legs reaching approximately to his ears, press three. If you wish a super tall, broad-shouldered, husky hunk with plenty of top hair, press four."

You press. "Beep," goes the phone.

"Thank you. You have reached Preference Four: tall, husky, with plenty of top hair. If you prefer your Country Man's top hair to be blond and wavy, press 5. If you prefer dark brown or black and curly, press 6."

Excitedly, you press twice, "Beep! Beep!"

"Thank you. Your Custom Country Man order is being processed with complementary parts to match. Please stay on the line for enhancement options."

"Thank you. Select enhancement options for your preferred Custom Country Man as follows:

"Tight Wranglers and plaid shirt—press one. Tight Wranglers and plain blue-denim shirt—press two. Tight Wranglers, leather vest, and white pirate-style shirt open at the neck showing chest hairs—press three."

"Beep! Beep! Beep!"

"Thank you. If you wish your Custom Country Man to be sent to you within the next ten working days by regular postal package service, press one. If you desire your Custom Country Man sent by UPS, press two. For Express Mail, press three. For special messenger, press four."

"BEEEEEP!"

"Thank you. Your Custom Country Man will reach you within the next two minutes. Custom Country Man Services appreciates your business. To end this call, press pound."

"Click." You hang up.

Heart galloping, you fling open the door and there he stands, your very own Custom Country Man, an unshaven, saggy belly, bald galoot, romantically attired in raggedy shirt and pants smeared with nasty stuff you don't want to know about.

"Greetings," you say to your spouse. "I see I've pushed the wrong buttons."

SECTION FORTY-ONE

Country Mouse, City Mouse

Wherein one offers advice to those newcomers to the country searching for their personal Haven, and wherein greenhorns experiencing their first tastes of rural life are highly entertaining to us local yokels.

Greenhorns to the rural west are people from all economic levels. Some fit right in, especially those from blue-collar backgrounds who already know how to work. They start a garden, get the kids into 4-H, join local civic groups, and learn *how to do country chores.*

Other newly arrived folks walk around wearing bewildered expressions. They tend to ask questions like, "What's a cattle guard?" Going to town for parts has no meaning for these neophytes to country living and "chores" is only a funny word. The following advice is free.

Chores do not allow time off just because it's Sunday or because it's too hot, too cold, or too tedious. If you've broken something such as a leg or an arm, you still have to do chores. (Hint: Wrap the cast in a plastic bag to keep hay and grain dust from creeping inside and itching like crazy.)

Whether male or female, relief from doing chores is not an option, though there are certain assists that can be developed. Such as having children. A pregnant woman, for example, can get out of chores for a day or two just by going to the hospital to give birth. A clever woman might contrive to go into false labor a couple of weeks ahead of schedule, thus earning extra time away from chores. Such an individual naturally takes along a pile of favorite reading material.

Farm and ranch chores can occasionally be Tom Sawyered by inviting a friend who has a kid or kids old enough to handle shovels, but not yet old enough to expect wages in any form other than cookies and milk and a ride on the tractor. This ploy can sometimes get the chicken house shoveled out. However, after one experience in chicken-house cleaning, the youngsters tend to smarten up and no kind of cookie bribery works.

To obtain a willing worker to clean stalls in the horse barn, locate a teen-aged girl who's wild about horses but has no equine of her own. Promise her use of Old Dobbin in exchange for shoveling. An added perk here is that Ms. Teen often brings along one or two girlfriends; therefore, it is advisable to have extra shovels and rakes handy.

A similar approach to chores can be employed in the elimination of pesky varmints (the ones that are not on somebody's endangered list). If your pasture is overrun with gophers, find a reliable teenage boy and let him target-practice. Like the chore-girl in the barn, the chore-boy in the pasture will likely have a buddy only too glad to help.

Warning to city-turned-country folks: Most chores are gender-related if you're male, but not if you're female. If the tractor breaks down while *Himself* is running it, *she* goes to town for parts. If the tractor breaks down while *Herself* is driving it, *she* still is required to go to town for parts—plus explain what she did wrong that created the problem.

Space limitation prevents a complete summary and description of rural chores, but here is one particular bit of advice to those City Folk who want to become Country Folk: Like getting dressed every day, doing chores is an absolute must. Nobody notices what goes into constantly keeping up with these everyday tasks—unless they're left undone or you show up naked.

To help the City Folk who may decide to vacation in Rural Land this summer, here are a few tips and definitions from the Country Hick Dictionary.

In the City, an agent is someone who keeps track of appointments and bookings.

In the Country, an agent works for the county and keeps track of weed-spraying.

In the City, a tuxedo is something you wear to formal occasions or on the job while waiting tables in a fancy restaurant.

In the Country, a tuxedo is what you name your new black-and-white foal.

In the City, spurge, larkspur, and knapweed are considered flowers for spring bouquets.

In the Country, it's a constant battle to keep the aforementioned pests under control using sheep and goats to eat the stuff or chemical sprays to discourage the poisonous plants completely.

In the City, some bathrooms come equipped with a bidet—a supplemental cleansing unit for certain parts of the human anatomy.

In the Country, whether it bidet or binight, do-it-yourself is the purification method of choice.

In the City, shoppers shop at Body Beautiful Boutiques.

In the Country, folks wait for a yard sale.

In the City, a cow is seen as a character in a nursery rhyme.

In the Country, a cow is the factory that produces food.

In the City, some view the eating of red meat as an abomination.

In the Country, no one eats *red* meat—they cook it first.

In the City, calves are the lower muscles in the legs just behind the shinbones.

In the Country, calves are what cows give birth to once a year.

In the City, a tube is a slang expression for television.

In the Country, you "tube" a bloated cow by inserting a piece of hose down her throat to let the gas out.

In the City, a spade is a black image on a playing card and, if you're playing bridge, it's the best "suit" to hold.

In the Country, a spade is an irrigating shovel, which you use to direct water-flow over grain and hayfields.

In the City, grass comes as a form of dried weed and can be rolled, smoked, and yes, even inhaled.

In the Country, grass is what is eaten by bovines before they become steak.

In the City, cutting a wide swath is akin to seeking lots of attention.

In the Country, cutting a swath is to mow down tall spring and summer growth using temperamental machinery.

In the City, a wheeled outfit that has a cab and an open box behind is called an SUV—sports utility vehicle—and is viewed as a status symbol.

In the Country, a similar vehicle is called a pickup and is viewed as a working tool, a transporter of livestock, and occasionally a second home if the spouse is mad at you.

In the City, a gate is often a decorative structure through, which you can pass without debarking from your vehicle.

In the Country, a gate is usually a barbed-wire unit you must get out of your pickup or off your horse to open, pass through, and *close it behind you!*

ODE TO CHORE COATS

(Tune: Clementine)

On a hook out on the back porch
Hangs my chore coat so divine,
It is made of canvas duck cloth,
It ain't purty, but it's mine.

Chorus—
Oh, my chore coat, oh my chore coat
Oh my chore coat so sublime,
You are ripped and torn and tattered,
Dreadful sorry, chore coat mine.

Wore you when I calved the heifers,
Got you covered up with slime,
Now you're stained and awful icky,
And you smell plumb unrefined.
Chorus

Wore you when I gathered cattle,
Through the swamp and 'cross the creek,
Dropped you in the muddy water,

You've been leaking for a week.
Chorus

Stretching fences in the springtime,
Tore a pocket on the wire,
And it flapped there quite forlornly,
Had to staple it with pliers.
Chorus

Branding calves, I accidentally
Burned the left sleeve most in half,
Turned around and burned the other,
Now they're both a perfect match.
Chorus

Now my chore coat's held together
With some duct tape and some twine,
Safety pins in place of buttons,
What a shame, oh, chore coat mine.
Chorus

Though you're old and nearly useless,
And you're in a sad decline,
Still I love you like no other,
I wear you now for *auld lang syne*.
Chorus

Ready for market in less time at big saving in winter feed

Hollow Tile farm buildings are always warm, dry, and easily ventilated, because of the two or more air spaces in each unit of their wind-tight walls.

Winter live-stock being prepared for the spring market will put on more fat, with materially less feed, in warm, dry, well-ventilated buildings than they will in cold, draughty quarters.

Hollow Tile Farm Buildings

The Most Economical Form of Permanent Construction

The first cost is practically the same as well-built frame farm buildings—and it is the last cost, for they are permanent and do not require continual repairs and paint. Their smooth, sanitary walls are rat and vermin proof and they offer a fire-resistance badly needed by every breeder.

Hollow Tile farm buildings are not experiments. They are profitable farm-building investments and pay big dividends.

If you are going to build, get our free book, "Hollow Tile Farm Buildings," and you will be convinced that Hollow Tile farm buildings are the most profitable.

Be sure that your local lumber or building-material dealer furnishes **MASTERTILE.** *Hollow Tile so trade-marked indicates material made in accordance with Association standards. It is your protection. Insist on getting it.*

THE HOLLOW BUILDING TILE ASSOCIATION
Representing America's Leading Manufacturers
BY THIS TRADE MARK—MASTERTILE—YOU SHALL KNOW IT
CONWAY BUILDING · CHICAGO

"BROOKLYN BRAND"

SULPHUR

COMMERCIAL SULPHUR, 99½% pure for spraying—insecticide purposes and potato blight

SUPERFINE COMMERCIAL SULPHUR 99½% pure for dusting purposes

FLOWERS OF SULPHUR 100% pure

Manufactured by

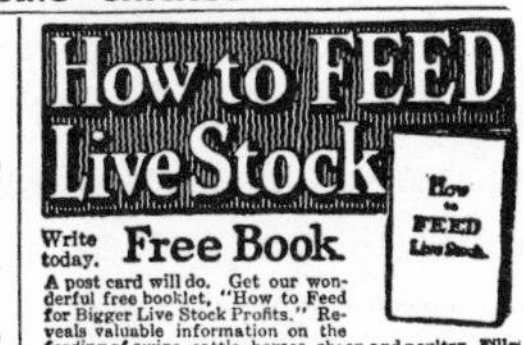

Write today. **Free Book**

A post card will do. Get our wonderful free booklet, "How to Feed for Bigger Live Stock Profits." Reveals valuable information on the feeding of swine, cattle, horses, sheep and poultry. Filled with interesting pictures. Also describes our wonderful course in Live Stock Feeding. Right methods of feed-

SECTION FORTY-TWO

Basic Farm and Ranch Attitudes

(A Ranch Is a Ranch Is a Ranch Is a Farm Is a Farm Is a Farm . . .)

Wherein one learns that regardless of the labor, the chapped and peeling complexion, the lack of elegance, and the dearth of shopping centers, one still counts oneself the luckiest person in the world.

No Country Woman has to worry or wonder about her "role." It's a partnership arrangement. You, your husband, and the kids work continuously to wrest from nature a decent living. As a Country Woman, you are sure of one important thing. You share the battle. You share good and bad. You share the terrific feeling when the farm or ranch has a good year. You share the thrill the time there's almost a hundred percent live calf crop. You also share the gloom the year the grasshoppers eat the corn and hail ruins the wheat just before harvest. One winter you share all the feeding chores with the kids when your husband fractures his leg.

If there's one thing a Country Woman is not, it's pampered. But it's an exultant feeling to know you're a part of every aspect of the business of living. As a Country Woman you know how many critters are due to calve or lamb or pig. You know when the haying is due to start, when the cattle have to be rounded up, and what to expect from the livestock buyers. Every member of the family is an important and responsible part of the operation.

Unlike the City Woman who associates with her spouse only during weekends and on social occasions, you live every day all day with your man. There's no such thing as keeping business and home life separate on a ranch or farm.

There's lots of room in the country. There's room for the Country Woman to have plenty of daily opportunity to admire her Country Man. By the time the City Man goes to work at eight or nine, your man has put in half a day's hard physical work before breakfast. He's a loveable character with a limitless capacity for hard labor. Soft spoken (except in the corrals under certain circumstances), he quietly manages all the toughest jobs around the place. He can ride, lift all that heavy stuff, drive machinery, deliver calves, throw bales, shear sheep, rope critters, shoe horses, make speeches at the local farmers' meetings, get involved with farm and ranch politics if necessary, figure the best nutritional balance for feeding livestock, do carpentering and electrical work, doctor animals, build corrals, and tote his kids patiently around with him, teaching them from his own experiences.

There's poetry and a soul-filling satisfaction for those who like working with land and animals. Some City Folks are misplaced farmers at heart and would take to country life like the proverbial ducks to water.

However, a working farm or ranch is not for the idealistic dreamer who depends on "magic thinking." Magic Thinkers tend to be unable to relate the application of the shovel to the ditch or the manure pile. They feel Nature and God will manage everything. Through mystical contemplation alone, the animals and the crops and the gardens will produce, and the glorious bounty will fall into the Magic Thinkers' open mouths. With all due respect to the Deity, Nature, and Contemplation, crops, animals, and gardens sure do better if you help 'em along. Ranchers and farmers face a constant, back-breaking, fiercely competitive fight against the weather, floods, drought, disease, and predators, followed closely by taxes and governmental regulations. The battle rages each year and each year it is never really won—at best it's a standoff.

However, be it a good year or a poor one, if you're lucky enough to live in the country, you're guaranteed to have one heck of a good time! There's nowhere else you'd rather be. Wherever you go, whatever you do on the ranch or the farm—toting a snack to the hay fields, trotting out to the corrals, collecting the eggs, or picking berries along the creek banks—you sure know you're alive and living in the country. The miracle of every day's small incidents and routine happenings constantly help you reckon you're a Country Woman.

For example, you know you live on a ranch or farm:

...when the increasing height and breadth of the manure pile gives you a warm feeling of pride.

...when you glance out the window to note your man marching up the path, a newborn, half-dead calf dangling over his shoulder, and you automatically jump to spread a blanket behind the heater stove.

...when other women brag about their household achievements and you brag about the three sets of triplet lambs.

. . . when your youngest son in kindergarten describes the birth of a calf during "show and tell" starting with, "First, a big bubble comes out the back. . . ."

. . . when you laugh wildly because town women call you to serve on assorted projects because "you don't work."

. . . when a city kid comes to visit and you overhear Number One son explaining the function of the little house out back with the two-holer seating.

. . . when King, the seventy-pound collie, hits a skunk, sneaks in the house, and jumps on the bed.

. . . when your husband sleeps on the couch all during his favorite program.

. . . when you open the refrigerator door and a vaccine bottle drops into the Jell-O.

. . . when your man pleads with you to drop everything and drive immediately to town. It's an emergency—the tractor needs a part.

. . . when it's five minutes before supper and you're scraping the leftover peas into the leftover potatoes, planning for soup and crackers. Just then you hear your Country Man on the back stoop cheerfully assuring a whole gaggle of extra men, "Sure, there's plenty! C'mon in."

. . . when you wash your face, put on lipstick and a fresh blouse, and your husband wants to know where you're going "all gussied up."

. . . when "dinner" means a ten-course meal served precisely at high noon.

. . . when everyone you know or ever did know thinks a perfect vacation is a visit to a farm or ranch—yours.

...when Peek-a-Boo, the pet sow, roots open the back door to the utility room, finds a full milk bucket and a basket of laundry, and proceeds to blend everything into porridge.

...when Lancelot, the banty rooster, likes to perch on your head and crow.

...when your hands and arms swell up for a week from picking buffalo berries.

...when you try to educate a city friend in the art of butchering her own chickens so as to economize and she turns green and leaves the room in the middle of your recital.

...when you accidentally inject yourself while vaccinating the cows and have to go see the vet instead of the MD.

...when you get drafted to milk the cow who has just come fresh—and your hands swell up like sausages.

...when your heart lurches, just like in books, at the breathtaking crystal beauty of the mountains against the spring sky.

...when you see the row of gray fence posts marching in measured formation across the winter snow.

...when the miracle of new grass in spring begins to push green spears among the winter brown and gray.

...when your favorite sow has fourteen nice, healthy piglets and she loves them all.

...when the hot summer nurtures each moment of the day in a cocoon of peace.

...when a wintertime chinook makes you feel as bouncy as a calf.

And you really know you're a Ranch or Farm Woman when your man puts his arm around you as you both lean over the corral fence watching the fat hogs eat up that expensive grain and he declares,"Happy Anniversary, you old Heifer!"

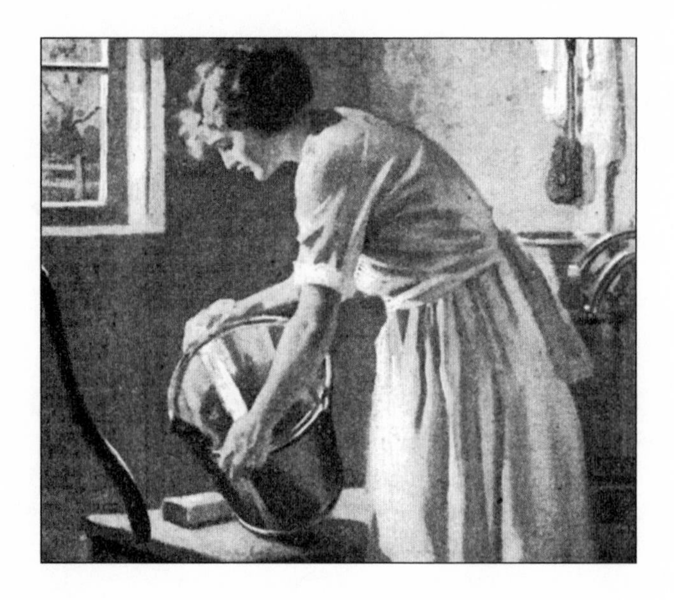

A WOMAN OF THE LAND

This is a tribute to all Country Women who have helped forge our American landscape.

Her name won't be in history books,
This woman of the land,
Her heart is where it wants to be,
Content with Heaven's plan.

And in the corridors of time,
Her way is counted true;
Enduring hardships, strong as rock,
She does what she must do.

Born when wire had begun
To stitch up prairie seams;
When homesteads patched the Western Land—
Quilts of hope-filled dreams.

A herder's wagon was her home
In sun or rain or wind;
She and Pa did outside work,
While Sis and Ma stayed in.

Her Pa hired out his greyhound dogs
To track the coyotes down,
With hounds in cages, he would drive
From ranch to farm to town.

Pa never settled down for long,
For when the bounty slowed,
They packed the wagons, hitched the teams,
And traveled down the road.

Sometimes Pa would trade for goods
Like shoes for her or Sis;
And once he got a violin,
Its song was like a kiss.

But Pa, he sold the fiddle for
Another purebred hound.
She watched it go without a word,
And cried without a sound.

They lived awhile at Charlo's house,
The Flathead tribal Chief;
A gentle man and kind, she said,
Who never showed his grief.

In hand-me-downs she walked for miles
To learn to read and write;
For school was just a sometime thing,
Wherever Pa would light.

But when she reached her sixteenth year,
She caught a cowboy's eye;
He asked her, would she marry him—
And bake him apple pie.

As partner, double-harnessed now,
She worked beside her man;
They saved and bought a modest spread,
And settled on their land.

She toiled in cold and snow and wet,
Or heat that scorched her bones;
No 'lectric lights to chase the dark,
No plumbing in their home.

She saw the land was good and strong,
And planned to run some sheep;
And when he called her fool, she said,
"Those woolies, I *will* keep."

They cared for cattle side by side,
Though sheep were hers alone;
But in the fall, the lamb checks paid
The interest on the loan.

Then Nature dealt another hand—
A child was due in May;
And though it didn't slow her much,
She did take off one day.

The baby boy was strong until
Pneumonia won the fight;
The child was buried on their land,
She battled grief at night.

Then, trailing cows, her pony fell
In crashing, crushing pile;
Her belly took the saddlehorn,
And town was thirty miles.

Her cowboy found her, took her in;
To God he made a plea.
They patched her up, her scars grew dim,
But children weren't to be.

She poured her spirit into work
On land that gave—and took;
She held no grudge and never cast
A single backward look.

As years slipped by in River Time,
Her cowboy lost his sight;
They sold the ranch and bought a place
More suited to his plight.

One day he started 'cross the road,
She shouted—caught her breath!
Her cowboy never saw the car
That dashed him to his death.

And now she ran the ranch alone,
And with her partner gone;
The only thing remaining true—
The land was there each dawn.

As time and strength began to wane,
She took another chance;
And leased a tiny piece of earth—
A widow-woman's ranch.

There's sheep and rabbits, goats for milk,
And hens for eggs or stew;
She doesn't wait for other folks
To tell her what to do.

She's planted trees to shade the house,
And sowed some grass for hay;
She irrigates her rocky patch,
Stays busy through the day.

Her hands are gnarled, her step is slow,
Yet when she's asked to town
By kindly Senior Center folks,
She always turns them down.

"I haven't time, I've chores," she says,
"They're what I aim to do.
My heart is where it wants to be,
My land will see me through."

Though her name won't be in history books,
And her range is less than grand;
Her heart is where it wants to be—
This *Woman of the Land.*

ABOUT THE AUTHOR

For Gwen Petersen, the seeds of country living were planted during youthful summertime visits to her grandparents' Illinois farm and by her mother who saw to it that they raised chickens in their city backyard. When Gwen married a rancher, her background as an occupational therapist for the bothered and bewildered helped her some in figuring out how to cope with country life. After thirty-five years of muddling through, she's starting to catch on. Today Gwen lives near Big Timber, Montana, where she raises miniature horses and works on her attitude. In addition, she continues her various writing pursuits, performs at cowboy-poetry gatherings, and ramrods the annual Sagebrush Writers Workshop.